Bearings

AMERICAN UNIVERSITY STUDIES

SERIES V
PHILOSOPHY

VOL. 210

PETER LANG
New York • Washington, D.C./Baltimore • Bern
Frankfurt • Berlin • Brussels • Vienna • Oxford

JOVAN BRKIĆ

Bearings

An Essay on Fundamental Philosophy

PETER LANG
New York • Washington, D.C./Baltimore • Bern
Frankfurt • Berlin • Brussels • Vienna • Oxford

Library of Congress Cataloging-in-Publication Data
Brkić, Jovan.
Bearings: An essay on fundamental philosophy / Jovan Brkić.
pages cm. — (American university studies. V, Philosophy; v. 210)
Includes bibliographical references.
1. Philosophy. I. Title.
B29.B73 100—dc23 2012038712
ISBN 978-1-4331-1176-1 (hardcover)
ISBN 978-1-4539-1036-8 (e-book)
ISSN 0739-6392

Bibliographic information published by **Die Deutsche Nationalbibliothek**.
Die Deutsche Nationalbibliothek lists this publication in the "Deutsche
Nationalbibliografie"; detailed bibliographic data is available
on the Internet at http://dnb.d-nb.de/.

The paper in this book meets the guidelines for permanence and durability
of the Committee on Production Guidelines for Book Longevity
of the Council of Library Resources.

© 2013 Peter Lang Publishing, Inc., New York
29 Broadway, 18th floor, New York, NY 10006
www.peterlang.com

All rights reserved.
Reprint or reproduction, even partially, in all forms such as microfilm,
xerography, microfiche, microcard, and offset strictly prohibited.

Printed in Germany

For Lee

Table of Contents

Foreword ix

Acknowledgments xv

Warm-Up 1

Definability 4

Neti Neti 7

Space-Time 10

The Known Universe 15

The Biosphere 18

The Separation Region 22

The Sociosphere 29

The Phenomenon of Family 39

The Phenomenon of State 42

The Phenomenon of Religion 49

The Phenomenon of Norms 53

Where's the Future? 71

Glossary 80

Foreword

The material contained within this book has come from what might be considered an unusual place. Dr. Brkić, the author of this scholarly and informative piece of work, has had long-term close associations with many truly gifted, brilliant people. Over many years, the group of friends would get together to exchange ideas and discuss topics of mutual intellectual curiosity.

This "Fellowship of the Mind" would meet in restaurants, clubhouses, homes, or just about anywhere they could sit with a cup of coffee, a glass of wine, or water and share knowledge on subjects of mutual interest. The synthesis of those discussions is contained within the pages of this book.

To this day we still gather over our preferred beverage and continue to explore realms of human knowledge and philosophy. We have all benefited greatly from these discussions. We feel that none of us as individuals could have gained the insights on topics of discussion without the knowledge and wisdom provided by others within our fellowship. In this case, truly, the total is far greater than the sum of the parts.

The "Fellowship of the Mind" consists of Dr. Jovan Brkić, professor of philosophy; Leland Swanson, Jr., business man; Robert Anderson, business process engineer, Steven Langer, M.D.; John Strandness; and Bill Hansen, attorney. Dr. Brkić and we of the "Fellowship of the Mind" hope you will benefit from reading this work as much as we have enjoyed contributing in our own unique way to its content.

There are several definitions for the word "bearings" — "direction or relative position" and "the manner in which one conducts or carries oneself" just to name two. In this book, both definitions are appropriate.

To find our "bearings" in life we need many different pieces of information, which provide coordinates or points of reference that, when taken together, provide guidance and direction, "bearings" for our journey through life.

Dr. Jovan Brkić is a singularly unique scholar and a man of applied learning who is a retired philosophy professor from the North Dakota State University. Born in Belgrade, Yugoslavia, in 1927, Dr. Brkić holds a doctor of philosophy degree from Columbia University, with extensive knowledge in linguistics, cosmology, mathematics, religion and sociology. Dr. Brkić has contributed a considerable volume of work in his field to scholarly periodicals and has presented numerous papers at academic meetings. He has published many well-received books, including *Norm and Order: An Investigation into Logic*; *Semantics and the Theory of Law and Morals*; and *Legal Reasoning: Semantic and Logical Analysis*.

A true renaissance man of great depth and breadth, Dr. Brkić has a unique gift to connect the dots between many diverse intellectual disciplines, providing a coherent philosophical body of knowledge that connects key elements of human understanding and endeavor.

Within this book Dr. Brkić explores the philosophical and logical connections between such varied topics as Space-Time, The Known Universe, Consciousness, Soul and Mind, and many others, providing the reader with a rich tapestry of intellectual continuity on diverse topics seldom found in one place. This book is rich in intellectual content and challenges us to think about the topics in ways that we may have not done before.

Bearings provides the reader a set of coordinates for helping determine personal direction and location, as well as a foundation for the synthesis of many intellectual disciplines, in ways that are both challenging and rewarding. Dr. Brkić provides another set of tools for understanding the complex set of interactions within the human social, religious, physical and moral conditions that make up the fabric of our lives.

This book is not intended to be a novel read sequentially from cover to cover. Rather, the book is intended to be read in a way that suits each particular reader's interest. You can pick up the book, open to any topic, and receive a complete and concise understanding of that singular topic. Each chapter within the book provides the reader another set of coordinates or "bearings" for locating our direction and position in the larger context of human understanding.

—Robert C. Anderson

Acknowledgments

The author would like to thank:

Dr. Catherine Cater for comments and suggestions.

The late Dr. George Graff for comments and suggestions.

Alexandra Griffin for editing and typing the first draft of the manuscript.

Vida Edmond for editing and typing the last pages of the manuscript.

The "Fellowship of the Mind": Lee Swanson, Bob Anderson, Dr. Stephen Langer, John Strandness, and Bill Hansen, for their friendship and support.

Warm-Up

Savants generally consider the age of the universe to be in the range of fourteen to twenty billion years, a finite number. If the velocity of light is finite and constant, as the savants believe, then by the laws of transfinite arithmetic a light signal run backwards would arrive at the beginning of the universe in a finite time. There it would have to halt, since no space-time can exist beyond the beginning of the universe. Consequently no science, nor thinking of any known kind including the artificial intelligence type, can be possible beyond the beginning of space-time. Some have claimed, however, that there is an Ultimate Source of All Existence beyond the beginning of the universe with its space-time.

A light signal departing the earth can be vectored in any one of 64,800 directions, If only intersections in whole degrees of latitude and longitude serve as launching points. Over long distances, the light signal is expected to encounter on average the same distribution and density of matter, regardless of the direction to which the light signal is vectored. The former property of the universe is called isotropy, the latter homogeneity. These properties, however, are only abstractions that cannot be corroborated beyond a doubt, needed though they are for generalizations about the universe.

If the beginning of the universe is unreachable in a finite space-time, implying that the space-time is infinite in the direction of the past, there still would have to be an Ultimate Source of all Being, or Everything, from which all existence emanates.

In either of these cases, the Ultimate Source of Everything has been conceived by some as a personal God, by others as an impersonal Supreme Being, by still others as an indefinable wholly Other. But no matter how it may be conceived, it is not accessible to any logical modes of definition, nor any description. In this respect, it is indeed transcendent in the Kantian sense.

How the Ultimate Source of Everything relates to the universe, or to universes if there are more than one, are metaphysical questions that are unanswerable through any rational or empirical means. However, there have been claims throughout history that such questions are answerable through other ways.

With the universe as it is, one can start talking seriously about its accessibility to the rational constructions and verifications of such by measurement and observation. This can be done in principle, without reservations if the universe is closed, it being understood that it is closed if its boundary can be reached from any point in finite space-time.

Purely rational and empirical procedures exclude valuational concepts perfection or beauty, indeed even evolution if it is loaded with presuppositions of "higher" and "lower" life forms.

With these provisos, it can be said that the least complex entities in the known universe are atomic and subatomic particles. Variants of the Quantum Theory, notwithstanding logical and verificational difficulties, seem to deal adequately with this level of existence.

As one reaches a biolevel of complexity, all known theories fail to streamline, and hence explain observed and verifiable facts of existence. Whether bioforms exist elsewhere in the universe is an open question. So are questions concerning the correlation between the physiological processes, notwithstanding the accumulation of an enormous amount of data and an equally huge number of claims suggesting that theoretical streamlining and organization of the processes are at hand.

If one subsumes the Special Theory of Relativity under the General Theory of Relativity and Newtonian Mechanics under the Special Theory of Relativity, and takes electromagnetic phenomena as covered by the Quantum Theory, then the relevant scientific theories are at hand for organizing and streamlining physiochemical macrophenomena. Also available for the confirmation of these theories is substantial, but by no means overwhelming, evidence.

Comprehension of the bearings depends on the language used to store and process information gathered in the universe. Central to all of this are definition and definability.

Definability

Identification of entities is made by terms occurring in natural or formal languages. Natural languages, at least on earth, are a product of their communities, and are learned initially by listening to communications between members of a language community. Formal languages, such as the language of mathematics, are constructed for specialized purposes, starting initially with a natural language. Although in principle one could use only a natural language for purposes of mathematics, it could hardly be possible to do so in practice. In fact, a mixture of natural and formal language elements is used in practice. Thus, formal languages are not context-free in practice, since natural languages also are not context-free. Formal languages do not, however, have prosodic features that natural languages do. Both formal and natural languages use terms to talk of entities that may or may not be definable.

Both natural and formal languages use function components to organize terms in context. Thus, the words of any language, natural or formal, are used to name or link words into terms and sentences.

If there exists a definition of a word, that word is definable. For example, by a noted theorem of Alfred Tarski, truth is not definable in the language in which all the truths are expressed. Moreover, by the combined results of Kurt Goedel, Alfred Tarski and Alonzo Church, it is not possible even in formal sciences to determine by universally applicable methods what is true and what is false. These results make it virtually impossible even in formal sciences to make claims to absolute truth.

None of the formal definitions are equally firm. Recursive definitions, for instance, are firmer than the mathematical context definitions. Dictionary definitions of natural languages are essentially natural language contextual definitions. Some are firmer than others. None are as firm as formal definitions. Dictionary writers use a variety of techniques to clue in the reader how to understand the word being defined. None of the defining techniques can deliver a conclusive interpretation of the word's meaning without vagueness, ambiguity or slipperiness.

In both formal and natural languages, the word being defined, the definiendum, usually is exhibited by bold face letters, the object of language. The words used to define the definiendum, the definiens, also are exhibited by special print, the metalanguage.

The definienda are either names, atomic or compound, or otherwise grammatical function words. The interpretation of the function words depends essentially on the linguistic context. The interpretation of names depends essentially on the kind of entity being named. Thus, the definiendum of the name "ten" is interpreted as the name of the number 10. It presents no interpretation problem. The definiendum of the name "Paris" is interpreted as a name of the city of Paris. The city of Paris is observable and subject to verification by empirical methods, including measurement, although by no means unproblematic. For example, it is impossible to determine with absolute precision its city limits or the airspace above it. All the same, the observability of the city of Paris ensures identifiableness of the city of Paris.

Far-reaching conclusions can be drawn from these facts. First, in formal sciences—logic, mathematics, and computer science—the definienda can be conclusively interpreted because the entities they name can be conclusively interpreted because the entities they name can be conclusively determined. Second, in physical sciences—physics, chemistry and biology—the definienda can be determined, though by no means unequivocally, because the entities they purport to name cannot be pinned down beyond a doubt.

The definienda purporting to name entities that cannot be formally or observationally identified, like "integrity," can be defined only through the linguistic context, the avenue used by dictionary writers. Because of this fact, even the existence of such entities can be vouched for only by the language context that includes personal and cultural contexts. One can convince oneself that this is indeed so by this test: "Integrity is a virtue" can express either a true or false statement. However, the nonsense "integrity is edible" expresses neither a true or false statement. One concludes, therefore, that the speakers of a given language must somehow be able to conjure to themselves what kind of entity "integrity" names by the very fact of being able to identify the correct and incorrect usage of such a word.

A far-reaching conclusion can now be made for the class of entities purportedly named by the definienda like "integrity," "justice," "honesty," "beauty," or "goodness." No formal ascertaining of the existence of such entities is possible, since no formal definition of the definienda is possible. Also impossible is the empirical assurance of the existence of such entities, since they cannot be directly observed. It remains open whether some indirect methods exist that could make them accessible to observation and hence ascertainment.

Neti Neti

Neti neti, meaning "neither this nor that," was used by Hindu sages to talk about the undifferentiable and ineffable nature of Brahman-Atman. Western medieval mystics used "via negativa" to convey the idea that God can be defined only in negative terms by saying what He is not rather than what He is. Therefore, no attributes expressible in any language can be properly associated with the Ultimate Ground of All, according to the Eastern and Western mystics.

It is interesting to note that the "negative" definitions like neti neti and via negativa can be raised to the standards of formal definitions. Thus, the complement of a set S is the set of elements that are not members of S. For example, if P is the set of positive integers, the complement of P is the set of negative integers. The via negativa definition of the Ultimate Ground of All then defines it as what cannot be seen, touched, heard, smelled and thought. The set of existing things and beings is therefore the complement to the Ultimate Ground of All.

The mystics of India, the Islamic Community, Europe and China claimed that although the Ultimate Ground of All is ineffable, it is accessible to the mystics' insight, acquired under very special circumstances and by very few.

Medieval theologians used "finitude" as an attribute of creatures and "infinitude" as an attribute of God. Modern cosmologists use Infinitude as a possible attribute of the Universe. However, both finitude and infinitude as used by both the medieval theologians and modern cosmologists can at best be defined through dictionary definitions.

The concepts of finiteness and infiniteness, in contradistinction to the usage of infinitude and Finitude, receive an elaborate treatment in formal sciences. For starters, there is a whole infinite hierarchy of infinite sets—it will suffice to deal with the sizes of sets only—starting with the smallest infinite set and progressing to ever larger ones.

The set of natural numbers is the smallest infinite set. It is definable by a recursive definition, considered the firmest kind of definition, by stating the initial, induction and closure conditions of recursive definitions. Thus:

    A. Zero is a natural number (initial condition);
    B. Whatever is a successor of a natural number is a natural number (induction condition);
    C. Nothing else is a natural number (closure condition).

Implicit in this definition is the fact that every natural number can be augmented in a step-by-step manner by the function "successor of . . ." Furthermore, every natural number defined in this way is unique. From the definition of the infinite set of natural numbers starting with the number zero, one can arrive at the number one as the successor of the number zero, and the number two as the successor of number one, etc.

A defining property of infinite sets is that a subset of an infinite set can be the same size as the infinite set. For example, a set of even natural numbers is a subset of natural numbers. Yet its size is the same as that of the set of natural numbers; that is, both sets have the same cardinal number denoting the size of the sets. The proof of this fact is elementary: Any set whose members correspond exactly with one member of another set and conversely are the same size. Consequently, since every element of the set of natural numbers corresponds exactly with one even number, and conversely, both sets have the same size. The set of natural numbers is enumerable or countable and every set having the same cardinal number as the set of natural numbers is enumerable.

Since the set of natural numbers is the smallest infinite set, a smaller set could be a subset of it, but could not be infinite. This leads to the definition of sets of finite sizes: a set is finite if it is not infinite and if it cannot be a proper subset of itself. The size of finite sets is denoted by finite cardinal members, the usual zero, one, two three, etc.

A set of integers is a subset of the set of real numbers, by not conversely. Hence, the cardinal number of the set of integers is smaller than the cardinal number of the set of real numbers, sometimes called the power of continuum. The set of real numbers is a subset of the set of mathematical functions, but not conversely. Therefore, the size of the continuum is smaller than the size of the set of mathematical functions.

Space-Time

Since ancient times, space and time were conceived as separate entities. It was Hermann Minkowski and Albert Einstein who promulgated the concept of space-time as a single entity. This conception of space-time in Relativity Theory is what essentially distinguishes it from Newtonian physics, which follows traditional usage.

Rene Descartes' discovery of how to construct a geometric model of arithmetic and conversely made possible the formulation of n-dimensional geometrical spaces through elementary logical procedures. Hence, the geometrical aspect of a four-dimensional space-time entity presents no problems. But the notion of a physical four-dimensional space-time is felt as unnatural. Yet, the opposite should be the case in view of the fact that no event can happen in space without it happening in time, and conversely.

The largest component of a formal or natural language is sentence-counting a finite sequence of strings as one long sentence (or formula). Largest components of sentences are strings (counting words as strings). Finally, the largest components of strings are letters and symbols in written language and phones in spoken language. Strings can name entities or express relations between entities. For example, the string "Paris is northwest of Rome" expresses a geographical relation between Paris and Rome. The strings "Peter is the father of Paul" and "y=f(x)" express the functional relations between *Peter* and *Paul* and the variables x and y. Natural languages have finite lexicons, but since they are open-ended they can be enlarged indefinitely. Formal languages work usually with infinite lexicons.

A semantic or mathematical model for a language is associated with a collection, the *universe*, which contains at least one element so that it is not empty. The *universe* of a mathematical model consists of individual entities and ordered sets of entities usually called *n-tuples*, some of which are ordered pairs, triples, etc. A given language is related to the corresponding model by an interpretation function that matches strings of words with appropriate elements and *n-tuples* of the *universe*. A *model* for a language is therefore an ordered pair, the first element of which is the universe; second, the interpretive function.

A theory in a given formal language is a set of sentences. A set of sentences with the same logical consequences as the theory constitutes the axioms of the theory. Hence, the set of axioms of a theory is the initial set of sentences that are taken as true and that characterize the initial words of the theory. Subsequent words of the theory can be introduced only by formal definition.

A sentence of a theory is true if there exists a universe and an interpretation function relating the components of the sentence to n-tuples of the universe. For example 2+2=4 is true because there exists the universe of integers, the addition function, the equality relation in the arithmetic model, and the interpretation function that relates the sentence 2+2=4 to the universe model. A theory is closed if every sentence of the theory is true, or is a consequence of a true sentence.

An element u of a given universe is definable if there exists a defining string in the corresponding language, such that u is the only element in the universe that satisfies the defining string.

It is clear from this that none of the theories of the physical sciences—in particular quantum theory, general relativity, or thermodynamics—can satisfy the stringent requirements of a formal theory. Indeed, physicists themselves make no such claims. Their only claims are that the physicists' theories, which are sometimes called models, must be logically consistent and testable by established methods of the given branch of science.

Sentences that are testable by scientific methods usually are presented in the form of algebraic equations, or equations of the differential and integral calculus. They could therefore function as formal definitions, provided they satisfy the requirements associated with formal defining. However, this is not the case. Thus, only sharper than usual dictionary definitions can be produced for scientific purposes. This is, in fact, what is claimed when the cliché "a more precise definition of . . ." concept is adduced. Notwithstanding the "softness" of definitions in science, a remarkable record of logical consistency and empirical confirmations has been achieved.

Generalizing from gathered empirical data constitutes a much more difficult and complicated problem. Empirical tests are applied in practice on a small subset from a very large open-ended set. Conclusions and findings based on a small sample are extrapolated to a large—usually assumed to be infinite—superset. Such generalizations are sometimes reliable, sometimes not, depending on the properties of tested objects.

The fundamental constants of physical sciences are the speed of light, the Planck constant, the electric constant, the gravitational constant, the magnetic constant, and the charge on an electron. All of them presuppose the existence of space and time.

The speed of light c and acceleration a can occur only over distances in space and durations in time. Consequently, space and time are the ultimate concepts of physical science.

Mathematical entities, in contradistinction to the physical ones, presuppose neither space nor time. This can be established quickly and simply. For instance, geometrical entities are usually introduced as loci of points satisfying certain conditions. But a geometrical point has no dimensions, whereas a physical point has four dimensions, counting time as a fourth dimension. Specifically, a geometrical point has a position though no magnitude; a physical point has both a position and a magnitude. Again, if light rays are fanned out from the center of the Earth in every direction, only a finite number of them can be fanned out at a given time. An infinite number, indeed a non-enumerably infinite number of geometrical curves arises from the geometrical center of the Earth. If one adds to them the fundamental propositions of general relativity, special relativity, quantum theory, and thermodynamics, it becomes clear that the fundamental concepts of physical sciences are interdependent, though not always interdefinable. For instance, from

$E=mc^2$, where the definiendum 'E' is defined by the definiens "mc^2", one can obtain $m=E/c^2$, and $c=$ *the square root of E/m*.

These equations show that mass, energy, and speed of light are interdefinable. The interdependence but not the interdefinability of entropy and time can be seen from a deceptively simple variant of Clausius' Second Law of Thermodynamics:

The entropy of a closed system increases with time.

The expanded version of this law reads:

Let S be entropy, let T be time, and let C be a closed thermodynamic system. Then:

S is greater than or equal to 0 if and only T is greater than or equal to 0.

It follows from this that the concepts of entropy and time are interdependent but not interdefinable, and that neither entropy nor time ever decreases.

Assuming that light rays are fanned out from the center of the earth, the question arises: how far can they traverse through the space-time coordinates? The Einstein equation in conjunction with other laws yields a number of answers. Since $E=mc^2$, one concludes that $c=$ *the square root of E/m*. If E/m represented an infinite quantity, the light travel could go on forever. In fact, some Indian sages claimed that the Source of all Energy, the Brahman-Atman, is inexhaustible. But this is not so if the law of entropy holds. For by the law of entropy, all processes in the universe, if the universe is a thermodynamically closed system, are subject to decreasing energy and increasing disorder, ending eventually in the death of the Universe. Now the Universe is at lease semi-closed in that it had a beginning. Furthermore, since the light rays consist of quanta of photons, they are finite quantities. But all finite quantities of energy are subject to the law of entropy.

If the universe is not closed in the direction of the future, the light rays could continue to travel forever. This, however, entails the future-directed infinity of time. That raises the question of what kind of infinity. Infinite distances in geometrical terms entail the power of the continuum. Indeed, the space-time continuum is addressed in the General Relativity Theory in that sense. Also, there are directional geometrical distances whose cardinal number is the power of the continuum. Not so the directed distances in the physical world because there is no geometrical continuum in the quantum world of atomic particles. One concludes, therefore, that no presently known theoretical software fits the hardware of the universe.

In view of all of this, it seems probable that the physical space-time is not even enumerably infinite. Gigantic steps will, of course, be made to compile more data organized and explained using more specific theories. However, there are no conclusive answers to be expected.

The firm point of departure for terrestrial beings is Earth. A finite number of directed distances from Earth are open in every direction, extending indefinitely through space-time with no end in sight.

The Known Universe

Until recently, atoms, intuited in ancient India and Greece, were believed to be the smallest particles of matter. Moreover, empirical evidence to back up the belief in the existence of atoms, and that their interacting forms molecular and larger compounds through electromagnetic forces, is of recent vintage.

During the past few decades, a whole new world of subatomic particles, together with the "weak" and "strong" radiation forces that make subatomic particles interact, was discovered.

Macroscopic entities, known from the earliest times, interact through gravitational forces. An important difference between gravitational forces and electromagnetic and radiational forces is the fact that shielding against the latter is possible, but impossible against the former. Another important difference, this time between gravitational and electromagnetic forces on the one hand and radiation forces on the other, is that the former are long-range while the latter are short-range. For instance, gravitational and electromagnetic forces manifest themselves in galaxies and radio waves, whereas weak and strong radiation forces materialize over very short nuclear distances.

The linkage of subatomic particles by radiation forces in constituting atoms represents the ultimate basis of material existence.

The succeeding level of existence is arrived at by compounding atoms into molecules and substances by electromagnetic forces.

Finally, gravitational forces participate in the linking and interacting of masses in any order of magnitude.

The basic ontology of the physical universe can now be listed from purely physical evidence: the movement is from entities of smallest magnitude to those of greatest magnitude. This is to a large extent in agreement with ontologies of ancient India and Greece. The important difference, however, is that ancient ontologies blend quantitative aspects of existence with those that are purely valuational. This means that more complex levels of existence are conceived as also more perfect ethically and esthetically. The movement through the levels of existence is hence conceived by ancient ontologies as a simultaneous development from smaller in magnitude and perfection to larger in magnitude and perfection. The approach proposed here, in contradistinction to that of the ancients, is to take out the "perfection scale" and retain the quantitative scale.

The most important case illustrating the difference between traditional ontologies and the one proposed is the Theory of Evolution. The Theory of Evolution explicitly and implicitly entails the idea of lower and higher forms of life, understood in the sense of "less valuable" and "more valuable" Thus, animal forms of life are lower than the human forms of life. This idea can and has been carried to the point of justifying the sacrifice of animals for the sake of human consumption and experimental purposes. No such justification can be made with the proposed ontology, for the purely quantitative ontology packs no valuational baggage.

Most interactions in the universe happen in a series of irreversible processes that are partially determined by an interaction of measurable thermodynamic entities: the internal energy of an isolated system, the work being done on or by the system, and the heat transfer to or from the system. The laws of the conservation of energy, entropy, and the zeroth law of thermodynamics relate to the thermodynamic entities. The conservation law equates the change of internal energy of the system to the work done on it and the heat gained or lost by it. The entropy law relates the change of entropy to the passing of time. Thus entropy, understood as a disorder or disintegration in irreversible processes is always on the increase. Lastly, the zeroth law of thermodynamics states that if one entity is in thermal equilibrium with another, and the latter entity is in thermal equilibrium with yet another entity, then the first entity is in thermal equilibrium with the last.

There is yet another factor by no means negligible in the physical levels of existence. That factor is what is nowadays called chaos. It denotes unpredictable, random behavior in systems modeling deterministic laws. Examples of such behavior are a turbulent flow of fluids, oscillations in electrical circuits, chemical reactions, dripping faucets, and unpredictability of the weather over longer periods. In some respects, Werner Heisenberg's Uncertainty Principle, by the very fact that it relaxes strictly deterministic laws on a quantum level, is also a witness to chaos in the physical universe.

Once the outline of the ontological framework of the physical universe is at hand, the question arises whether intelligent existence within it is possible. If by intelligent existence some sort of computer-like existence is understood, the answer must be affirmative. For a computer consisting of some sort of hardware, with memory and processing units and software to pack the program for the operating system, is capable of doing what intelligent entities do. Since computers can process information, do calculations, self-examinations, and self-programming, it follows that intelligent existence does not presuppose biological existence. So a biological level of existence is not needed for the existence of intellectual entities.

Is it possible that a known Universe and everything in it has been programmed? Quite likely. Could it have been the Ultimate Source of All where the programming originated? It is logically possible. Is it possible for the Programmer to intervene in the operations of the program? It is logically possible but empirically unlikely. Could there be an appeal from within the programmed existence to the Programmer? It is logically possible but empirically unlikely.

The Biosphere

The Earth is a component of the solar system that in turn is a component of a multi-billion collection of stars, the Milky Way Galaxy. The sun and other stars of the Milky Way Galaxy exhibit no features that would distinguish them in their physical and chemical composition from other stars. Even the Earth's atmosphere, so far as its physical and chemical properties are concerned, is not much different from any other celestial body: one encounters radiational, electromagnetic and thermodynamic forces in action on Earth as much as anywhere else in the universe. What makes Earth so strikingly different, according to our current scientific knowledge, is the biosphere that developed on it: the complex molecular structures that evolved, eventually into elementary life forms and ultimately into living organisms capable of initiating and moving complex biochemical processes.

Organisms are open thermodynamic system, that is, they exchange matter and energy with the environment. The movement of the biochemical processes is governed primarily by the laws of thermodynamics, particularly the law of entropy. This law directs them toward their peaks and ultimately their cessation: Death in ordinary language. Hence, the overall patterns of living things on Earth are clear: the biochemical processes that initiate and terminate living things allow little leeway for intervention by agents and forces outside of the living thing's biochemical program.

The estimated creation of the solar system is approximately four and a half billion years ago. The increase of oxygen on Earth from two to twenty percent happened about a billion years ago: a necessary requirement for the existence of living organisms. The appearance of living organisms between six and five hundred million years ago on Earth appears to be a unique event in the Universe: on the foundations of the physical-chemical structures, high-rise structures of immense complexity and built-in operational programs have been developed.

Since the biosphere is unique in the discovered part of the Universe, any generalizations from it to undiscovered parts of the Universe are at best tenuous. The reasons for the tenuousness of the conclusions by extrapolation are logical and empirical. Assuming that the logic is applied correctly, one still faces the choice between the intuitionistic or classical logic. The choice depends on the philosophy. If intuitionistic logic is chosen, that logic is considered a branch of mathematics. Regarding matters involving finite sets, the difference between intuitionistic logic and mathematics and classical studies does not matter.

If one chooses classical logic and included mathematics, then proofs involving infinities do not carry the restrictions that proofs using intuitionistic mathematics and logic carry.

The problems involving generalizations from empirical data arise in connection with identification, demarcation, and hence measurement of empirical objects where oftentimes not even the firmness of dictionary definitions can be achieved. One instructive illustration of such problems can be seen by observing a wheat leaf under an electron microscope. The leaf surface magnified one thousand times the power of ordinary microscopes simply does not look the same as when magnified ten thousand times. When magnified one hundred thousand times, it is no longer recognizable as a leaf surface to the untrained eye. Question: does one see here the same object under different magnifications, or many different objects? Another question: how large is the leaf surface? The magnification at one hundred thousand times would certainly include things that are not visible at lower magnifications, and therefore would make the leaf surface look larger. If one were to reply to this that the experts should be able to resolve these problems, then the question of different interpretations by experts would arise. Thus, instead of being helpful, experts would make matters more difficult because of the problem of interpretation.

A large number of attributes appertains to concrete sets of objects in the biosphere. They can be sorted out into three classes with respect to generalizations by extrapolation. First, if the attributes of a potentially infinite set of objects are "firm" in the sense that extrapolations from any object in the set encounter no exceptions, then generalizations to the whole set would result in the laws of nature that seem to hold for the whole set. Second, if the attributes of a potentially infinite set of objects apply on the basis of a sample set, with a certain percentage within a numerical range, then generalizations can be made only within that percentage range.

One can talk in such a case only of statistical correlations, sometimes even of statistical laws of nature. Third, if the attributes of a potentially infinite set of objects are such that not even percentages can be determined, then generalizations can be made only by intuitive guesses. Such guesses cannot be described as "laws of nature." The first class of attributes appertains solely to some sets of objects in the biosphere; the second to other sets; and the third still to others. There are, however, sets of objects in the biosphere to which more than one class of attributes appertains.

The overall composition of the biosphere up to the psychosocial level is well known, including animal existence, although an enormous number of details remain to be discovered. However, as the psychosocial level of the biosphere is approached, the problems of definition and the language to effect them increase in greater numbers.

Thus, learning and memory, standard research subjects of the neurosciences, are taken to be associated with the biochemical processes, and therefore are subject to experimental verification. But is it memory and learning that are verified by experiments, or are the observed phenomena in experiments simply labeled 'memory' and 'learning'? What can justify labeling biochemical processes in plants and animals *learning* and *memory*? Are not such labels anthropomorphisms pure and simple? Would not for the purposes of science any other labels including mathematical symbols do just as well? It follows that such a procedure is not legitimate without restrictions, depending on the attributes defining the given sets of organisms. In the case of humans, the ability to communicate verbally, as one of the defining attributes of humans, is a decisive factor in deeming the labels 'memory' and 'learning' applicable to the whole set of humans.

Let a scientist set up a learning and memory experiment with a human subject. Let the scientist connect the brain of the experimental subject to required instruments, take required biosamples and proceed to "teach" the subject to identify several buildings on a picture. Let the subject be able to associate "taught" names with the buildings. Let the scientist thereupon analyze accessed biodata and locate the part of the brain where memory is stored and identify biochemical and electromagnetic processes in the brain associated with the "learning." The experiment is repeatable with the same or different subjects. Let the scientist then generalize the findings by extrapolation into the whole set of humans. No anthropomorphism by attaching human attributes to non-humans happens here. The only issue remaining in cases like this is how far is it legitimate to extrapolate from the findings coming from a human onto the whole, potentially enumerably infinite, set of humans. Therefore, this kind of extrapolation holds true generally in science.

The Separation Region

Memory, feelings, emotions, volition, and consciousness cannot be conceived and certainly not talked about without language. Although language capability is a human individual's attribute it is, as a means of communication, a societal phenomenon. Since there is no biochemical or biophysical machinery to relate individual humans, it follows that language transcends the biosphere. It is, of course, true that oftentimes animals also live in groups and communicate by means appropriate to the species. But such means of communication can in no way be considered comparable to human language. The transition from the biosphere to the psychosphere is therefore marked by the transition from the biological language capability to the acquisition of a particular natural language. Hence, through the acquisition of a particular language, an individual human being becomes a member of a particular society of humans. However, since all languages are intertranslatable with more or less delicacy, membership in a particular language group includes membership in mankind.

It was Noam Chomsky who drew attention to the fact that fluent native speakers of a given language can identify correct and incorrect speech structures to which they have never been exposed and hence could never have been taught to them. Actually, Chomsky could have claimed much more: native as well as non-native speakers of a natural language can identify correct and incorrect applications of words purporting to name individual entities or sets of them. For example, a minimal English language background suffices to identify "the garlic abused the onion" as an incorrect language usage and "garlic is a plant" as a correct one. Chomsky's point can be further strengthened by noticing that native speakers who have no physical handicap learn the sound features of their language without fail, something they cannot do with another language beyond a certain age.

Behavioristically oriented psychologists raised legitimate objections to the use of mentalistic concepts like "mind" and "soul," in that entities presumably named by such words are not identifiable and hence not amenable to empirical investigation. Indeed, such entities are accessible only through dictionary definitions; that is, through the linguistic context. This means that such words cannot be but vague and polysemous. The basic technique for curing vagueness is contouring the definienda to make them more precise; the basic technique for disambiguating the definienda is contextual analysis. The limits to arbitrary usage and interpretation of any natural language elements are set by the language community. Attempts to transgress these limits result in frivolous verbiage, and, if carried to an extreme, in a total breakdown of communication.

An instructive example of what happens when the plasticity of ordinary language is tampered with by radical procedures is to be found in the history of psychology. Believing that the word "soul" is not right for psychology, the nineteenth century psychologists replaced it by the word "mind." The critics of this method then dubbed psychology "the science of the soul without the soul." In the twentieth century, the word "mind" was deemed inappropriate for purposes of psychology, and thus was replaced by the word "behavior." Those who went to fix the problem in this manner missed an elementary semantic fact: "behavior" is a singular term purporting to name a unique entity. When it eventually became obvious that a mistake was made, the word could have been dropped, as is the case with "soul" and "mind," or a conversion of a singular term to a general term attempted. The latter exit from the mistake was taken: the singular term "behavior" became a general term. Singular terms do not pluralize, whereas general ones do. Hence "behaviors" was forced into English as a plural form and "behavior" as a singular form, whereas previously only "behavior" as a singular form existed. The lesson from psychology is clear: ordinary language cleansing, especially by non-linguists, is not the way to deal with abstract definienda that are inaccessible through formal definitions or observation. This is essentially the reason why social sciences cannot be "empirical" in the manner in which natural sciences are.

It is possible nowadays to associate an individual's introspection with its external behavior using language even under laboratory conditions. Language is therefore a bridge that can connect behavioral patterns at the atomic, molecular, and organismic levels with internal mental states. It does, however, much more than that. It also relates individuals with their introspective states and external behaviors with other individuals, creating social groups of various sizes and functions.

Dictionaries classify "memory" as a noun and describe its definiendum as the capacity of retaining impressions and recalling and recognizing previous experiences. The one sure test that fluent speakers of a language, in this case, English, "understand" the word "memory" can be immediately made: the sentence "Have some memory for breakfast" is anomalous in English as well as in translations into other languages. The sentence "I have a vague memory of my childhood" is correct in English and any other natural language translation.

The word "memory" has universal semantic features. These remain stable, even with metaphorical trimming added to it. There is only this exception: amnestic disorders and dementias, in particular aphasia, can impair the ability of fluent natural language speakers to identify accurately the correct and incorrect usages of a word. However, once sure language footing is abandoned, one encounters endless theories about memory and other mentalistic concepts, making it difficult to decide how much credibility any one of them merits. It is certain that there can be no memory without a nervous system, which includes neurons and tissues for conveying information between sensory cells and organs; the central nervous system with the brain; and the spinal cord and the peripheral nervous system. Add this to the electromagnetic processes of the brain and the body and one obtains what is in many ways a more complicated "hardware" system than that of a computer system designed by humans. It appears likely that the usual computer hardware mimics in some ways that of animal organisms. Since memory entails consciousness, it is quite possible that the two together are analogs of computer software.

An enormous amount of information concerning biochemical and electromagnetic processes of this hardware has been accumulated, and more will be coming. It is possible now to scan various parts of the nervous system and relate them to what researchers call "memory." It is, however, at this juncture that the scientific approach becomes questionable on logical grounds. Biochemical and electromagnetic processes are physical entities; memory is not. Without a credible bridge connecting them, external observations cannot be used to muse over internal experiences. Memory as an internal experience exhibits features like degrees of vagueness, confusion, forgetfulness, "indelibility," etc. None of these are attributes of biochemical and electromagnetic processes. It therefore follows that a reduction of memory as an individual's internal experience to biochemical and electromagnetic processes simply is not possible. What is possible is the relating of the former to the latter.

Feeling and Emotion

Dictionary definitions of the synonyms "feeling" and "emotion" associate physiological processes with the psychological experiences in a much more plausible manner than with the memory. Thus, the usual dictionary definition describes the definiendum of "emotion" as an affective experience of joy, sorrow, hate, fear, and love that is usually accompanied by physiological changes such as increased heartbeat, respiration, crying, and shaking.

Again, a sure test that fluent speakers of a natural language "know" what feeling and emotions are is manifested by their ability to identify correct and incorrect usages of these words. Thus, the following sentences exhibit respective incorrect and correct usages:

Love is a ghost.
Love is an emotion.
Hate is the opposite of whiskey.
Hate is the opposite of love.

Moreover, all four sentences show that the features attributed to the definienda are universal. Again, as in the case of memory, an enormous number of details concerning the observed phenomena associated with feeling and emotion are known, but many of them are unknown. However, of greater importance than the mere details is to identify the bridges that connect purely biochemical and biophysical phenomena to what in ordinary experience are considered mental phenomena. Also of paramount importance is to establish degrees of intensity of feelings and emotions by some methods of measurement that remain to be discovered, as well as the extent and character of storing feelings and emotions in memory.

Motive and Volition

Dictionaries define "motive" as something that prompts a person to act in a certain way or that determines volition. The dictionary definition of volition is "an act of willing or choosing." Dictionary definitions of both motive and volition entail at least a possibility of overt behavior. But overt behavior can only be affected through the nervous system. Its components are the central nervous system and the peripheral nervous system. The central nervous system's components are the brain and the spinal cord. They are presumably the operating system of volitional behavior. However, the participation of the peripheral nervous system, including the automatic nervous system, cannot be excluded as a factor in motivational and volitional behavior. The operation of the nervous system is fueled by energy generated by burning glucose. Thus quantum, biochemical and presumably psychological processes occur simultaneously and interconnectedly. Hence, it is impossible observationally or even theoretically to find a break in the sequence and simultaneously parallel such processes, making sure that an observed behavioral act or activity is motivated and voluntary on one hand and unmotivated and involuntary on the other. It is instructive to notice in this connection that in law and morals, claims are made that actions can be adjudicated as motivation and voluntariness.

Consciousness

The dictionary definition of the abstract definiendum of consciousness says that it is awareness of one's existence, thoughts, feelings, senses, and surroundings. This simply amounts in mathematical terms to an intersection of the finite memory contents and a given accumulation of points in time. One could say that the main difference between memory and consciousness is that consciousness is stratifiable by means of "more" and "less," the limit of "less" being unconsciousness, whereas memory is not. However, even in the case of memory one can talk of failing and fading memory in contrast to vivid memory. There is an enormous accumulation of data regarding the neurosystem and how its malfunction affects consciousness. But so far, as in the case of other mentalistic words, there is nothing else but the reports of the subjects to connect the conscious and the unconscious states of the subjects with the data collected through external observation. The situation gets even more complicated when dreaming states are added. If the subject recalls dreams, it follows that the subject was in some state of consciousness. In fact, ancient Hindu thinkers postulated just that: there are gradations of conscious states all the way into deep sleep. The gradation of consciousness from "fully conscious" and "less conscious" all the way to "unconscious" suggests that voluntary versus involuntary behavior is likewise subject to gradation, as it depends on the state of consciousness.

Soul and Mind

Soul and mind with analogous variants in other languages denote universally conceived entities. Thus, soul is conceived as a unifying principle of non-physical life, separate from the body, believed by many to be immortal, and mind is conceived as a comprehensive entity that contains mental states, emotions, and mental activities. The ability of respective language communities to identify correct and incorrect usages of "mind" and "soul" establishes their conceptual existence beyond a doubt. Doubts arise when attempts to relate them to their constituent elements and the elements to the biochemical and electrical processes of the nervous system are made. Some connections are readily available, but most cannot be established in any dependable manner. One can imagine the body as an analog of the computer's hardware and the mind as the body's software and operating system. There certainly are similarities. But one has to bear in mind that computers did not make humans in their own image; rather, humans made the computers in their image.

Thus, the logical concept of similarity is the basis of analogical reasoning that is involved when comparing "artificial intelligence" in computers with "natural intelligence" in humans. Indeed, the concept of similarity is the basis for much more than analogical reasoning. Therefore, generalization from samples to any whole natural kind, based on similarity, can only be made by assumption, backed up by limited factual evidence. Thus, conclusions about pain, suffering, motives, feelings, reasonings, and emotions are all based on assumptions of similarity with fellow humans and, sometimes, animals.

There certainly is a similarity between computers and humans in some respects, but not enough to draw conclusions about the body-mind relationship. A relation between body and mental states and activities does exist. But what it precisely is cannot be determined on the basis of the present knowledge; indeed, probably never. The reason for this is that while the definiendum of the body can be pinned down firmly by observational evidence, the definiendum for the soul and mind can be understood only by the natural language context.

There is no sharp separation line between the biosphere and the psychosphere in individual human beings, as one observes them apart from any associations to other humans. A sharp separation line occurs only at the border where the biosphere ends and the purely human sociosphere emerges. Thus, when no biological signification, not even atomic and molecular longevity if not immortality, is attributed to the concepts of soul and mind, they are both legitimate as well as meaningful in the sociosphere. With the possible exception of natural parents and children, there is no biochemical and electrical connection between humans in society. Therefore, forces other than quantum, electromagnetic and thermodynamic govern relations between humans. The fundamental questions are: What are these forces and how do they operate?

The Sociosphere

The initial objects of social inquiry are individual human beings situated on respective coordinates in the four-dimensional Minkowski-Einstein space-time. Since each one of them can occupy only one position with respect to each of the coordinate axis, the relations of each individual to any other individual state are these: an individual can exist simultaneously with another individual at different places or at the same places but at different times. It follows that the Minkowski-Einstein space-time coordinates order the set of humans in terms of the simultaneity or precedence in time, and different places in terms of space. An ordered set of this kind, though a necessary condition for the existence of human societies, is by no means also a sufficient condition. A sufficient condition can only be furnished by transphysical components.

Aristotle's position was that entities are characterized by essential and accidental attributes. Essential attributes are those without which an entity could not be what it is. Accidental attributes may or may not pertain to an entity; that is, their presence or absence does not make entities what they are. For example, a rose can be red, white, or yellow. The attributes of being red, white, or yellow are accidental attributes. But the botanical attributes that define roses as species of plants are essential attributes.

Aristotle's approach to attributes is useful for dictionary purposes, the only possible defining entities in the sociosphere. The ability to speak and understand a language in the sense that "so and so speaks such and such a language" is an essential attribute of every normal human being, "normal" being understood as not suffering from physical or neuropsychiatric disorders. Therefore, the predicate "so and so speaks such and such a language" defines a society of humans communicating in that particular language.

What kind of entity is society? What are the predicates that express essential attributes that define society? First, one can single out three kinds of entities that are definable by predicates expressing respective attributes. The first kind of entity definable by predicates like "so and so is a natural number," or "such and such is a ghost," or "so and so is justice" is abstract. The second kind of entity is concrete, perceivable by sensory organs, and hence definable by descriptive predicates. The third kind of entity presupposes the existence of concrete objects, yet is not itself concrete. Such are justice, beauty, right, and wrong. One can think in this context of the old Platonic doctrine of participation. Plato saw the problem clearly, but his proposed solution of the problem was inadequate.

If a given sort of entities is such that each one of them can in principle be inspected or otherwise positively identified, they belong to a closed set of entities. Such, for example, is a set of natural numbers or a set of cities. If this is not possible, then the set is open. The set of all celestial bodies, if the universe is infinite, for example, is an open set. Therefore, what can be conclusively learned about a sort of entities depends on the nature of the defining predicates, the accessibility to sensory perception, and the closure properties of the set.

The fundamental essential attribute then that the entity called society has and by which it is defined is the linguistic ability that comprises communicative, descriptive, and interpretive abilities. Facts, data, and evidence do not exist in nature, only the phenomena accessible to sensory organs. It is by virtue of language that societies create facts, data, and evidence. Hence, society and its products are basically abstract entities sometimes associated with concrete substrata.

Thus, take any society conceived as a set of zero, one, or many members. If it has zero members in it, it is an empty set; it does not exist. If it has one or many members, it exists at a given moment as one or many interrelated bodies, hence as entities within concrete substrata. Their superstructures are mental and hence in part societal creations. The same society looked at from the perspective of the past exists only on the basis of interpreted historical facts. The same society, one again, looked at from the perspective of the future exists only in the imagination of the contemplator. In any case, a given society exists in the past and in the future only as an abstract entity.

Social products are of several sorts: manufactured items, services, and varieties of cultural creations. All of them depend ultimately on the presumed valuation by human beings. Everything is intertwined in what W.V. Quine and J.S. Ullian called the Web of Belief. Human actions and reactions are impossible to imagine without presuming that such and such is true and so and so happened. If somebody is held responsible for something, it has to be presumed that that somebody "knew" what was supposed to be done or not done and had the ability to act accordingly. Since most items in anybody's Web of Belief presuppose assumptions that are sometimes true but also sometimes false, the tenuousness of conclusions, and even observations, is obvious even in the case where concrete entities are involved. It is anybody's guess how much more so this holds in the case of abstract identities.

The initial and the most comprehensive language is the natural language. Formal languages, in practice an interlarding of natural languages with symbols, are simply extensions of natural languages. Hence, natural language is the ultimate repository of all the truths and falsehoods that can be assembled by verbal and non-verbal communication. Suppose one takes the words of English as an arbitrary natural language and enriches them with mathematical symbols as additional words. There are in principle an infinite number of words in this language. Suppose one classifies all such words that link words to form compound names, phrases, and sentences. For example, the word "five" names the number 5, the word "city" names the set of entities definable by certain predicates, "the last Ottoman Sultan" is a compound name identifying a historical entity. Examples of another kind, as in "snow is white," "five is a number," "iron is a gas," and "London is northwest of Paris," exhibit sentences that are composed by linking names and phrases.

There are three types of sentences that are definable by respective fundamental attributes. First, declarative sentences that express true or false statements; second, interrogative sentences that express answerable or unanswerable questions; third, imperative sentences that express performable or unperformable commands. For instance, "five is an odd number" and "three is less than two" express true and false statements. "Is four divisible by two?" and "how does one square a circle?" express answerable and unanswerable questions. "Wash your hands" and "become immortal" express performable and unperformable commands.

The correct application of attributes, the correct naming of entities, and the proper use of linking words to compose phrases and sentences determine whether one is talking sense or nonsense. The determination of whether something is true or false, answerable or unanswerable, performable or unperformable constitutes the decision problem of all spheres of human knowledge and activity. There is no effective general solution to this problem. There is, however, a general though not effective approach to this decision problem. The approach was best exhibited in practice by Plato and theoretically by Aristotle and Chaïm Perelman. The heart of the approach is a structured argument and discussion to convince respected audiences of the truth, answerability, and performability of what is under consideration.

Infinite subsets of the set of all sentences expressing statements can be singled out to check which consequences follow from the initial sets. If the initial sets express true statements, so do their consequences. Therefore, whether or not a given piece of reasoning expressible by means of declarative sentences is correct or not can be expressed in the initial sentences. It is, however, possible to analyze declarative, interrogative, and imperative sentences with respect to clarity, polysemy, and vagueness in order to determine their suitability as the initial sentences of a specific type of reasoning. Assuming this fact, possible modes of reasoning involve issues of truth, the properties of statements expressed by declarative sentences, the issues of answerability in principle involved in questions expressed by interrogative sentences, and the issues of performability involved in commands expressed by imperative sentences.

Elementary arithmetic, sometimes called recursive arithmetic, is considered the core of modern formal sciences. Some of the most profound achievements of the formal sciences have been made through the language of recursive arithmetic. But it can be shown that even in the formal sciences not all sentences expressing true statements are provable. Thus, there is no generally applicable algorithm for deciding what is true or false, even in formal sciences. This conclusion from formal sciences applies to an infinite model of reality that includes concepts and physical objects as entities. It follows that no universally applicable method for deciding whether an attribute expressed by the predicate "so and so is true" can be associated with an entity or not. Furthermore, even the definition of truth cannot be formulated in the same language as the sentences formulated expressing true statements.

Immanuel Kant maintained that transcendental questions like "does God exist?" cannot be answered in an existential context. The results from recursive arithmetic extended to include all the entities in the universe show that not even all existential questions are decidable in the existential context. However, in limited contexts many questions are answerable in the sense that correctly drawn consequences always follow from stated or assumed premises. This is true in matters of fact as well, but with the proviso that deductive correctness is a necessary but not sufficient condition for true conclusions. A sufficient condition requires the premises to be empirically provable.

Induction, including statistics, is indispensable in reasoning about matters of fact. In fact, inductive inference is not a separate mode of reasoning but an application of deductive inference to empirical matters, with no or inadequate closure properties. In practice, only objects in finite sets of small sizes are accessible to observation. Objects in large finite sets of concrete entities, though accessible to observation in principle, are not so in practice. Infinite sets of concrete entities, if such exist, are not observable in practice or in principle. Yet any kind of induction, including by statistical inference, is essentially over open sets with no possibility of closure through extrapolation.

Modes of Social Inquiry

A few samples from mathematics show what kinds of extrapolations are possible by means of the ultimate standards of reasoning:

- A. Let C be a fixed point with coordinates $x = a$ and $y = b$ in the Cartesian coordinate system. Then the locus of points P at a distance r from C is a circle whose center is C and whose radius is r. This conditional definition of a circle is generalizable to an enumerably infinite number of circles under the stated conditions.

B. Let F be a field of complex numbers. Then every polynomial equation has a root in F. Again, this statement of the Fundamental Theorem of algebra is generalizable without fail to any complex number and any polynomial equation.

C. Every *even number* except *two* can be represented as the *sum* of *two primes*.

To prove that this generalization, known as Goldbach's hypothesis, is correct, one would have to inspect every even number in the enumerably infinite set of even numbers. But this is impossible. To prove that no such generalization is correct, one would have to identify an even number that cannot be represented as a sum of two primes. Nobody has yet succeeded in doing this. Thus, not only the possibility of generalizing this case of Goldbach's hypothesis is in question but even the validity of the law of the excluded middle. The law of the excluded middle, which states that of the two contradictory statements one must be true and the other false, is indispensable in everyday life, as well as in legal reasoning.

D. Prime numbers *three, five, seven, eleven, thirteen,* etc. occur in the form of pairs p and p plus two. Thus, *three equals one plus two, five equals three plus two, seven equals five plus two, eleven equals (seven plus two) plus two, thirteen equals eleven plus two,* etc. To prove this, one would have to inspect the whole enumerably infinite set of pairs of primes in the stated form. But this is impossible. It is conceivable that some way could be found to prove the generalization, but so far none is available.

There are in general two kinds of laws in science that represent generalizations through extrapolations from finite to potentially infinite sets of objects. One kind is formulated by means of attributes that allow no exceptions. These laws become invalidated if a single exception appears. Such are in particular the laws of classical mechanics. The other kind allows exceptions within significant percentages.

These include quantum mechanics, thermodynamics, and biology laws.

It is of particular interest to notice that none of the generalizations in sciences carries such a high degree of certainty as Goldbach's hypothesis. There is also the belief that one can generalize from primes occurring in pairs of the form p and p *plus two* to the whole enumerably infinite set of primes. The reasons for this are again the kinds of attributes defining entities in the formal sciences and those in the empirical sciences. Therefore, the degrees of certainty in generalizing by extrapolation are essentially dependent on the kinds of attributes characterizing the entities generalized upon. Indeed, dependence on the kind of attributes is the thread that all generalizations in and out of science depend upon, provided that deductive inferences are correctly made.

Paraphrasing Leibnitz' law, one can say that any two or more entities are exactly identical if they share every attribute. Hence, two or more entities are totally different exactly if they have no attribute in common. One can then define similarity as sharing some, but not all, attributes by any two or more entities. The classification of entities into diverse classes is accomplished by establishing the relation of similarity between any two members of the class. This means that there are certain attributes that any two members of the class share. For example, a single attribute expressed by the predicate "x interbreeds with y" suffices for biologists to establish similarity between any two animals and classify them as belonging to the basic biological class, the species. Indeed, any generalization over entities presupposes the relation of similarity; that is, the sharing of some attributes.

If any entities can be shown to be similar with respect to certain relevant attributes, and hence to form a class of entities, then analogical inference can be applied to the whole class. Analogical inference is an essential tool of experimental science as it is in law and everyday life. For example, the vaccine testing results on a sample of animals can be extrapolated to the whole species and then even to another species, the humans, provided that the relevant conditions are satisfied.

Presumed similarity and analogical reasoning play an essential role in trying to understand how a human, or even an animal, "feels."

They also play an essential role in sympathy and empathy. The ability to empathize is essential to health care practitioners, possibly useful in limited ways to law enforcement personnel, and in general, in everyday human intercourse. However, there is no reliable way to determine which attributes, if any, are being used in empathizing, and hence, whether analogical inferences have been properly drawn. Therefore, a claim to "understand" a human being through empathy can be anywhere between a solid understanding of another human being to a total misunderstanding of another human being's feelings, emotions, and motives.

An important mode in legal reasoning is the weighing of evidence. It is comprehensive in that it can employ logic, methods of proof used in science, and a variety of methods for evaluating written and oral testimony. The weighing of evidence is in effect applying "common sense" by creating a context for grading the credibility of evidence in terms of "higher" and "lower." An example from biological theory will show how the process of weighing evidence can work. One biological theory postulates that "natural selection" plays a key role in the hierarchy of biological existence. The antonym of "natural selection" is "artificial selection." Thus, a wheat breeder can select a particular variety of wheat to grow or even to manipulate the wheat seeds genetically in order to develop the desired variety. Since artificial selection of a wheat variety occurs in this manner, one has to wonder if there can be any "natural" selection of a wheat variety other than wild wheat.

Several other terms are of interest in this context: "natural theology," "natural virtue," and "natural law." One can wonder whether any one of these exists, but they are the names of straightforward concepts. It is not so in the case of "natural selection." "Selection" implicitly restricts the choosing to humans. So the ability to select entities attributed to a biological process means anthropomorphism. For the sake of argument, "selection" can be used metaphorically in the theory of natural selection. Then the metaphorical use of "natural selection" allows organisms self-reflection in the process. It follows that wild wheat selects itself for the "survival of the fittest" along with other plants of the fields.

One can, however, ignore the absurdity of the term "natural selection" as an unfortunate label and consider the evidence in the chains of biological forms. The evidence of similarity between organisms is very strong indeed. They share many essential attributes. The paleontological evidence also is strong, but very scanty. It is by no means sufficient to indicate conclusive links in the postulated hierarchical chain of life forms. What is completely missing is evidence of any kind of natural or artificial selection. However, neither is there evidence that there was none. Therefore, whether "natural selection" as a mainstay of the theory of evolution is rejected or accepted makes no difference. It then follows that the Theory of Evolution does not merit the status of a serious scientific theory.

Every society on Earth exists in a four-dimensional space-time. Three dimensions of the space-time locate a society latitudinally, longitudinally, and vertically on Earth. The fourth dimension locates it durationally through the past, the present and the future. Every society has a beginning in the past, existence in the present, and eventual cessation of existence in the future. The blending of chaotic, purposeful, and orderly contours of every society's existence into a descriptive picture is history.

The biological substratum is the basis of every society's existence in the present and the sine qua non of every society's future existence, but no longer of any society's past existence. Therefore, the initial facts of any historiography are biological in nature, but are not the only facts. Others are wars, types of political and economic organization, and abstract values that seem to motivate overt behavior but are otherwise intangible.

It was the belief of Leopold von Ranke that history ought to be written as it actually happened. However, this can at best be a guideline to an objective approach in history writing rather than an achievable goal. Indeed, mere arrangement of facts in history into clusters that might be called historical evidence allows a vast array of different interpretations even without the injection of the historian's attitudes. However, in spite of such problems, von Ranke's guideline is valid; in fact, it is the sole criterion that distinguishes real history from enterprises like orchestrated propaganda and axe grinding. Not all facts are relevant for inclusion into historical evidence. For assessing their relevance, they have to be read within a context based on historical movements and trends that help in selecting a proper interpretation of historical evidence. Furthermore, historical movements, trends, and waves should be thought of as watcher phenomena with development possibilities in several directions, rather than some inevitable process, be it of cyclical or dynamic nature. The cyclical view of history in its Greek, Hindu, or Oswald Spengler's version has to be rejected because, although it is plausible in view of some historical evidence, other significant evidence makes this view impossible. The dynamic view of history, forcefully propounded by G.W.F. Hegel and Karl Marx, has to be rejected for the same reasons.

It is instructive to compare medical diagnosis and historical interpretation. Medical diagnosis weighs the physiological and psychological evidence with a view to deciding upon appropriate therapy. If it were not objective, it would be worthless or even harmful to the patient. But once the diagnosis is completed, the therapy is value driven. The value is the good of the patient. Historical evidence is weighed in the same manner as medical evidence. Any intrusion of value judgments to make facts or people look good or bad, any intrusion of uncontrolled emotional factors can only result in fiction, good or bad. Once the historical evidence has been properly organized and presented through an objective interpretation, the historian's work is done. The medical diagnosis, the physician's analog of historical interpretation, is only the prerequisite for therapy.

The proper subjects of history are the social phenomena, past and present, with a possible outlook to future ones. The future ones, if objectivity in history is to be preserved, must be no more than the reading of trends and tendencies.

The Phenomenon of Family

Biological species reproduce sexually or asexually. This means that sexuality is not necessary for reproduction. Furthermore, even in the case of sexual reproduction, mating is not necessary. This leads to the conclusion that although the relationship of a male and a female together with their progeny is usually conceived as a basis of a nuclear family, it is by no means necessary for the existence of a family as a social concept. In view of polygamous, polyandrous, and monogamous marriages—or setups like the one among Australian aborigines where members of sororities and fraternities get together for mating but otherwise live apart—family is to be conceived as a socially created entity whose forms and organization are not determined biologically.

If one considers a joint family where a clan and tribe are extended families, a picture of a purely social conception of family as a usual organization of patriarchal societies emerges. This organization is based on monogamous marriages in fact if not always in practice and custom. This is so because of the biological distribution of females and males in human societies. Hence, although polygamy and polyandry may be allowed by custom, they are usually impossible on a large scale in reality.

The evidence that the organization of humans in social units existed in some form everywhere on Earth is conclusive, as it is conclusive too that no particular form was necessary, nor a valuational sorting out of better or worse ones possible. It is not possible to pin down the causes, biological or social, that favor any particular form of organization. But once a particular form became prevalent, the necessary means to maintain and operate such a unit had to be found. First, ways of superordination, subordination, and coordination between members of a social unit had to be devised, as did methods of social control. Historical evidence for contouring the dictionary definition's definiendum of "family" seems solid, indeed more so than for many entities in the physical universe, but there is no sufficient evidence for backing up any particular conception for the organization of family.

Particular instruments of social control vary from one patriarchal society to another, though all of them are geared toward securing the preservation of the economic and social type of organization in existence. (The use of terms like "geared" is proper in the sociosphere just as the physicalist language is proper in the physical sphere.) The primary means of social control in patriarchal societies is custom, and in more complex ones customary law. The primary methods of enforcing custom and customary law are usually some form of shaming or disgracing the violators of norms, and honoring and glorifying those who uphold them. Physical coercion is used as a secondary means, though not uniformly. Imprisonment of transgressors in contemporary forms was unknown, but other forms of corporal punishment were used. Capital punishment was non-existent, unless one counts blood vengeance as a form of capital punishment.

The absence of punitiveness in patriarchal societies contrasts sharply with contemporary societies, which have devised complex legal and extra legal machinery to administer punishment. Likewise patriarchal religion, art, and music contrast sharply with those of contemporary societies. The heart of religion of patriarchal societies is the cult. It is essentially societal in nature: it plays the main role at birth, rites of passage, ceremonies, and death. The rites at birth and death at first blush appear as biological events, until one realizes that neither exists as biological entities. Birth is only a link in the chain of biological events, and death is the terminal link of the chain, at which point the law of entropy reduces biological processes to chemical ones. Therefore, there is neither birth nor death in the biosphere. Birth and death are societal demarcations of the biological process that are linked to the phenomenon of the cult.

If one starts with extant phenomena, the phenomenon of religious expression inevitably leads to the question of whether religion is purely social in nature, or whether humans have religious dispositions built into their neural system. If the same question is raised with regard to art and music, which are intimately blended with religious expression, the answer is affirmative. Art and music are based biologically, and hence religious expression must be related to the human biological make-up. Therefore, since the artistic, musical, and religious expression belong to the biological programming of human beings, religious expression is both biological and societal.

In social inquiry, one starts with extant phenomena, then fans out longitudinally into the past to identify events leading to present phenomena, and into the future to espy present aspirations and activities leading to future consequences. Lastly, one looks around latitudinally to contour and delineate the context of the past, present, and future phenomena. Casual explanations and statistical correlations are more often than not impossible and the road to historical figments of imagination wide open. Argumentation and careful weighing of evidence are almost the only protection from landing into fiction making.

With these provisos, one can anticipate that the joint family, the clan and the tribe as presently constituted will be relegated to history with no prospects for the future. The nuclear family, with its traditional forms of enculturation, will also be relegated to history along with the patriarchal forms of art, music, and the cult. Some form of the nuclear family will likely survive, but with recognized relationships to fit in with the total urbanization of the globe and the ubiquitous presence of coercive forces and surveillance in contemporary societies.

The Phenomenon of State

The Concept of a State

The definition of a state allows for the contouring of the definiendum of a state as an entity, constituting a coercive order whose present condition issues forth from its past, and whose present endeavors are energized by aspirations and goals of the future. Such a broadly conceived definiendum allows packing within its contours the conceptions of tribal, city, nation, and universal state. It provides for many ways of organizing any one of these types of state to fit extant states or purely theoretical conceptions.

Every state, regardless of type, has to be organized in terms of superordination, subordination, and coordination of relationships between its members. It follows that each state has to have a setup that implicitly enunciates the permissible relations between its members, and the means by which only permissible relations will be upheld. The former represents the policy of the state expressed through its law; the latter constitutes the machinery of the state's government.

Organization of a State

The policy of a state enunciates how its government machinery is expected to operate and what it is supposed to do. If the policy is that of an extant state, then one has a setup of a real state. Otherwise, the setup is a fictionalized design intended as a proposal of presumably an ideal state. One can leave out of the account fictionalized designs of state, like Plato's Republic, if one's objective is understanding what makes actual states tick. However, one cannot leave out designs like the Marxist model that resulted in serious attempts to make unworkable conceptions of state work. Such examples are instructive for showing what real life consequences of unworkable fictions can be.

The organic law represents the core of a state's policy in historically known states. It occurs in the form of customary law, both are written and unwritten and, in the forms of statutory law and judicial precedence. It is usually considered the most important branch of law, especially in the form of constitutional law.

Criminal law is in some ways an appendix of organic law in the sense that it spells out procedures and sanctions against the violators of organic law provisions. Likewise, administrative law can be considered an appendix of organic law, although its importance, due to sheer necessity in complex contemporary states, assumes a vastly greater role in the day to day affairs of individuals and state organs than constitutional law.

The core of private law, the regulation of relations between members of state, is the law of contracts. Commercial law, family law, insurance law and so on are simply specializations of the law of contracts.

Organic and private law spell out the policy of a state. However, there are no historical cases where the policy became realized to an extent where a realization could be anticipated with any definite certainty. Only tentative judgments of effective, less effective, and ineffective law enforcement are possible. If one considers morality as an informal backing up of law, as one must in Christian societies, not even this much can be claimed for the effectiveness of law enforcement.

The separation of governmental powers into judicial, legislative, and administrative did not exist in almost all of history. The customary law played the role of a legislature almost to the exclusion of rulers and legislative bodies throughout almost all of history. Even when the rulers did legislate, the legislation was minimal, as in the case of Asoka in India or in the case of Roman emperors in Europe, Asia Minor and North Africa. In the Islamic countries, rulers could not legislate for the Muslims at all, for whom the law was Sharia, which in turn was based on the Holy Quran. Even in Europe, where kings were supreme judges, administrators and legislators, their role as legislators was circumscribed by customary law, inherited Roman law and Canon law, and thus minimal.

A revolution in more than one way occurred with the proclamation of the separation of governmental powers and the principle of equality before the law by the American and French revolutions. The separation of governmental powers into the judicial, legislative, and executive branches was never complete; nor was the principle of equal treatment before the law, either de jure or de facto, ever fully realized. All the same, the very proclamation of these principles was in many ways more important than the compliance with their requirements itself. Of great pragmatic significance, made necessary by both revolutions, was the creation of governmental bureaucracies unprecedented in history, with a possible exception of China. Indeed, wherever large concentrations of human and material resources happened, so did the bureaucracy to administer them. It was therefore inevitable that under the contemporary circumstances of industrialization and urbanization of the globe, there also rose huge governmental and corporate bureaucracies. Carrying out governmental operations by surefooted but lethargic bureaucracies always implies modifications, stylizations, and distortions of original operating policies. This is true for any governmental function, be it the enforcement of policy, distribution of welfare, or development of projects. In fact, in more than one way rulers—elected, hereditary or self-imposed—have become no more than the key cogs of the governmental machinery. Witness how often the governmental machinery has been used to force out of office, or worse, top functionaries of assorted governments. The so-called rule of law, or nobody is above the law, simply confirms the de facto situation that the law itself is the product of the political process.

Policies and their administration in modern states, especially the powerful ones, have immense and to a large extent unpredictable consequences for their own populations and for those of other states. One can see a familiar pattern of such events in the following stereotyped form. After going through a process of decision making, a major power formulates a policy that will affect its own population as well as a host of other states and their populations either positively, negatively, or both. The policy will eventually have to be implemented by severe measures that the civilian and military bureaucracy will be expected to carry out by all means necessary in order to achieve the objectives of the policy. The populace of the country responsible for the policy may eventually be able to change it through protest or vote. The populaces of other countries, especially the minor ones, usually have no or very little impact on the policy initiated by another country, which affects them in sundry and deleterious ways.

The questions that arise are these: First, does the populace of the state initiating the policy and executing it have any obligation to the victims of the policy? Second, do the victims carry any responsibility for their contribution to their misfortune? Third, how is the responsibility to be distributed in the state initiating and executing the policy? The decision on the policy and its execution resembles in some ways the execution of a death sentence with distributed duties so that nobody personally kills the victim. A fundamental legal and moral principle requires that there shall be no responsibility without authority. The victims can hardly be said to have had authority over what happened to them. But neither can the victimizers be said to have had a controlling authority, since it was distributed and served in stages down the ladder of the hierarchically organized bureaucracy.

The conclusion to be drawn is that the fundamental legal and moral principles professed by past and present states are in many cases irrelevant and in some cases deleterious to victims and victimizers alike. If one adds the plethora of unenforced and unenforceable laws, one has to wonder how the behavior of contemporary humans is to be controlled. A cursory look indicates that this is done in part by enforced laws, in part by customs, and in part by ignoring the unenforceable legal and moral principles in highly complex societies.

The decisive differences in the phenomenon of the traditional state and the contemporary one is due not as much to religion, culture, or language as it is to the economic factors, the industrialization, and the urbanization. The family, the village, and the community were the basis of historical states, the governmental machinery simply being superimposed upon them. In contemporary states, urbanization and industrialization have shattered the family and almost destroyed local communities in many parts of the globe. Traditional legal and moral values, cherished by major and minor religions of the world, are either irrelevant or outright negative. The following examples illustrate the point:

1. Competent surgeons may subscribe to whichever moral and legal values they like. The incompetent ones may too subscribe to whichever values they like. Competent surgeons will benefit their patients; incompetent surgeons may harm them. Hence, it does not matter to which moral and legal values the surgeons subscribe. Moral and legal values are irrelevant in the surgeons' practice, unless one counts competence as a moral or legal value, which nobody has proposed yet.

2. The same can be said for any profession, but especially for governmental functionaries in a contemporary state. Thus, competence overrides any moral or legal values that may be at play in governance of public or private enterprises. This also was the case to some extent in the past, but the harm that can result due to incompetence in contemporary situations bears no comparison to historically known examples.

Economic arrangements and technical skills in states have always exhibited similarities regardless of nationality, culture, and language differences. Contemporary states exhibit the same similarities but with one fundamental difference: modern technology and economic and business activities exhibit an overwhelming reliance on scientific know-how rather than traditionally acquired skills. The world has been unified through economic and business activities in spite of political, religious, and national differences. Contemporary science and technology are poised to enable humans to travel through the solar system similarly to the way seafaring know-how enabled Europeans to travel on Earth. Since the perpetuation of the species is biologically dependent, so is securing the economic wherewithal to do it. The space and the resources of the solar system and beyond are literally the only opportunity for further expansion and growth of the human race.

The separation line in art, literature, and music between pre-urbanized and pre-industrialized and urbanized and industrialized states and societies is not as noticeable as it is in other fields. The need of humans to create and enjoy art, literature, and music is presumably rooted in their biological makeup rather than their social setup, if one discounts valuational factors like "good art," "profound poetry," or "great music." However, one cannot minimize the impact of radio, television, and motion pictures in universalizing and thus adding, shaping, and integrating whatever comes from biological factors. No state had existed in the past without religious or ideological support.

The earliest forms of religion were tribal. Henotheism, and lastly universal religions, emerged with the more complex states. Many contemporary states fit some or all of these patterns of religion; some none. There is a large variety of religious phenomena represented through diverse religions making attempts to generalize meaningless or outright false ideas. This much, however, can be said without running the risk of making insupportable generalizations: every state has had at least one type of religion to officiate at the rites of birth, marriage, and death. Contemporary states show some departures compared to preceding states with respect to religion. This is due partly because of the intermingling of religions and secularization of contemporary societies, but also overwhelmingly because of their globalization and urbanization of these.

Contemporary states are withering in more than one way, though not in the Marxian sense. Interdependence of shared resources and living space on Earth necessitate voluntary restrictions on claims to sovereignty. An exception would be if one state harbors serious plans to resolve the interdependence by some sort of extermination or severe containment of other states, which then would lead to endless trains of disruptions, dislocations, and victimization.

These facts will force a radical departure in interstate relations compared to historically known cases of universal states: the Chinese Empires, the Hindu Empires, the Roman Empire including its Byzantine form, the Arab Khalifates, the Ottoman Empire, the Russian Empire, the British Empire, and the demised Soviet State. The exception is the United States, which in some ways fits the historically known pattern but differs in others. Whatever the case, the demise of the Soviet Union, representing a new mix of world order and world chaos that remotely resembles the Athenian hegemony, has emerged. The role of Athenian democracy is now being played by the United States. However, the similarity merely echoes the Athenian model because industrialization and the depletion of the Earth's resources will force a severe braking on further development and expansion, unless the avenues into the solar system, its resources and space, are opened up.

The Phenomenon of Religion

The phenomenon of religion exhibits existence in relation to transcendence through sets of beliefs that even atheistic religions like Buddhism and Jainism share. The Transcendent One, the Ultimate Source of Everything that exists, cannot be reached from the platform of existence. The Hindu sages and the mystics of every culture knew this. David Hume and Immanuel Kant showed that proof of God's existence based on the platform of physical existence cannot establish a causal connection between existence and transcendence. But neither Hume nor Kant showed that there is no Ultimate Source of Everything. Indeed, the complexity and the mix of order and chaos in the existing universe lead to the idea of an Ultimate Source of All Being. That is all that can be shown. How the Ultimate Source of Everything might be intuited and conceived is beyond any demonstration. The mystics of all climes and times are in agreement on this: discursive reasoning and empirical proof cannot break the barriers of existence in order to reach transcendence.

Worship consists of prayers and rites. Worship, together with a body of beliefs, constitutes the cult. It is the cult that all religions have in common, regardless of the biological makeup of humans participating in the cult, the clime, the time, the language, and the culture. There are, however, special attributes that are shared by some religions and ideologies. Students of religion and ideology missed it, but it was not missed by a novelist. In her novel *Lagum*, Svetlana Velmar-Janković identified the attribute by asking the following question: what makes the Communists coming from different walks of life, with different personalities, look and behave alike? What makes them capable of killing their friends and relatives? The novelist's answer is: the Idea.

The Idea flowing from the Gospel and the Dogma, the Holy Quran, the Communist Manifesto, and Mein Kampf is not to be found in Hinduism, Buddhism, Taoism, Confucianism, or even in Shintoism and the Greco-Roman polytheism, and certainly not in tribal religions. The Idea embodies the claim to Truth, postulating explicitly or implicitly that only those who profess the Truth can be saved, short of special dispensation. Since doing good is part and parcel of the Truth, in the idea the desire to save everybody by winning them over to adopt the Truth of the Idea generates the dynamism lacking in non-Idea religions. This dynamism leads to the missionary zeal and activities that eventually lead to conversion efforts, resulting in direct and indirect pressure, and sometimes in violent force, to accept the soteriologic Truth of the Idea.

The Idea is initially embodied in charismatic leaders. Indeed, the originators of Idea religions and ideologies were, in the language of religion, mystics, in the language of psychiatry, sufferers of mood disorders, and in the language of law, reformers. Idea religions, though universalist, must not be confused with universalist non-Idea religions. Idea religions, to which the now defunct communist ideology and the extreme pathological case of National Socialism must be added, have their martyrs in the language of religion and victims in the secular language. There are, however, no victims without the victimizers; nor are there victimizers without the victims. Indeed, the victims oftentimes become the victimizers, energized by the need for vengeance and punishment for suffered wrongs, and in subtler ways by the "thirst for justice."

The followers of Idea religions are usually powered by an overwhelming faith in soteriological bearers of charisma, and are imbued with the missionary zeal to accomplish the salvation work expected of them. The requisite behavior and an intolerant attitude toward those outside the loop of the faithful then become the product of the psychological tune-up of the faithful. There are, of course, the behavior norms in non-Idea religions, but they are essentially tailored toward achieving one's own salvation, aimed toward individualization and not mass action. Not so in the Idea religions that are aimed toward absorbing the personality traits into a uniform attitude, geared toward mass action and mass look.

The peculiar phenomena of the 20th century are the now defunct ideologies of Communism, National Socialism, and Fascism. All three exhibited certain features through which they could be categorized as Idea religions, and were able to generate uniformalizing attitudes in their followers. All had among their leaders individuals with assorted psychiatric disorders, and were powered by enormous hatred against the victim groups, without the animosities arising from the usual socio-psychological connections. There are, however, significant differences between them. Fascism was a relatively minor phenomenon in terms of the harm to its victims, or the promise to the victimizers. Most importantly, it laid no universalist claims.

National Socialism and Communism, through the sheer number of victims and victimizers, are in a class by themselves in world history. The National-Socialist movement, created by a demonstrable psychopath, raised questions that cannot yet be answered in a conclusive manner. What makes humans accept psychopaths for their leaders and become galvanized by the kind of rhetoric delivered by them? National Socialism, by the very claim of racial superiority and exclusivity, is not universalist. Not so Communism, although it too is not devoid of psychopathic leaders. It has had, however, a mix of both relatively normal and abnormal leaders. All the same, it professed true universalism without regard to race and nationality, yet also exclusivity to social origins and social classes.

Democracy and Nationalism exhibit initially the appearance of an Idea ideology. But, as one takes a closer look at Democracy, one notices the absence of the cult, the victims and the victimizers, the charismatic leaders, and a sparse scattering of the psychopathic leaders. Nationalism does generate victims and victimizers, the cult is often associated with it, and it is not devoid of charismatic and psychopathic leaders. All the same, it lacks the Idea, the basic ingredients in Christianity, Islam, Communism, and National Socialism.

As one contemplates the phenomenon of religion and ideology, its ubiquity, the institutions associated with it, and the societal and psychological impact it has, one must reach the inescapable conclusion: religion or ideology are part and parcel of human nature originally programmed into human beings. One must also reach another inescapable conclusion, as one contemplates the diversity of religions, conflicting beliefs and practices: the original Programmer did not favor any one of them; from the phenomenon of religion, there is no "true religion."

The Phenomenon of Norms

The phenomenon of norms is as ubiquitous through climes and times, in some ways even more so, as the phenomenon of religion. No known society, large or small, ever existed that did not have norms prescribing implicitly what the right conduct is, what the proper thing to do is, and the right way of doing things. It is, however, a fact that actual behavior is never in full agreement with what the norms say it should be. Thus arises the problem of norm deviation: in general, wherever and whenever customary law is in effect, deviation is exceptional.

Since tribal, and in general, patriarchal societies are controlled by customary law, deviation is also provided for by customary law sanctions. Sanctions are usually some form of corporal and psychological punishment that does not include imprisonment and death, unless one counts death through blood vengeance. However, death through blood vengeance is avoidable through reconciliation. The situation becomes quite complicated in "advanced societies" in which customary laws as well as norms generated by legislative, administrative, and judicial processes are applied. This is especially true of societies whose norms were generated through the blending of customary law, Christian moral teachings, governmental processes and philosophers.

For "advanced societies" one has first to establish which of the sometimes conflicting norms are really accepted, how a norm is interpreted, to what extent it is realized, and what the deviations from the norms are. Strangely enough, the major differences between the patriarchal and "advanced societies" is not in the norms themselves, nor in the methods of enforcement, but in the severity of enforcement. First, psychological methods of enforcement in "advanced societies" usually carry lifelong stigmata and disadvantages to those against whom they are meted. Second, corporal punishment in the form of beating, or imprisonment, or death are not only common but are demanded and expected in "advanced societies."

The biochemical substratum is the basis of the psychological makeup of humans. But what that makeup exactly is nobody really knows. There is an abundance of personality theories, but that is all that can be predicted from them. None of them can be proved to be either right or wrong. This satisfies Karl Popper's criterion for sorting out scientific from non-scientific theories. It follows that there is no proven way of conclusively interpreting the behavior of any human being. The problems of understanding humans grow exponentially as they congregate into larger groups. It follows that no theories of personality and human groupings can satisfy the criteria required of theories in mathematical and physical sciences. There are, however, facts in individual and group human existence that are demonstrably verifiable, sometimes more so than the facts in nature. The problem is that the verified facts of human existence are not generalizable into proportions that can function as the laws of human existence. Thus, generalized laws of human societal and psychological functioning are not possible, but reasoning based on solid circumspectly studied facts is.

Humans relate to each other space-temporally and socially. In either case, relations between them are ordered by spatiotemporal circumstances and the norms of conduct. The rules of conduct, expressed by the societal norms, presuppose that humans can exercise self-control over socially relevant behavior, that is, their conduct. This presupposition is to a large extent wrong. The realization of the wrongfulness of this presupposition becomes poignantly evident at the demarcation lines of the two World Wars. The two World Wars showed how tenuous, or even non-existent such controls are in the behavior of individuals, and even more so in the behavior of collectives of individuals. It cannot be argued that self-destruction and the destruction of others in those wars was the conduct of sane, rational and controlled behavior.

One has to distinguish sharply the generalizations and subsumptions falling under normative concepts, arrived at on the basis of empirical considerations, and those culled from writings under titles like "ethics," "moral philosophy," "moral theology," etc. The former represent attempts at real generalizations oftentimes arrived at by misreading the facts or making logical mistakes. All the same, such generalizations can at least claim to be based on the realities of human life. The latter are fiction, superbly presented, as in the Dialogues of Plato, or not so superbly presented, as in Kant's writings. The claim is often made about the great influence exerted by this fiction. The claim is totally false beyond the confines of the West and, within the confines of the West, it does not extend beyond the tiny intellectual elite.

There is no separation of custom, law and morality in patriarchal societies. As the more complicated social forms emerge with the ensuing economic, intellectual, aristocratic and clerical stratification, the crystallization of norms and values oftentimes emerges in the form of codes prescribing, proscribing, and idealizing certain kinds of behavior. Patriarchal as well as complex societies generalize norms and values into normative concepts that include, in Western terms, legal, moral, etiquette, and esthetic values.

The most inclusive generalization of values comes from Hinduism through the concepts of Dharma and Karma. Dharma generalizes ideas that make the world run right, and hence, in Western terms, entails religious, legal and moral duties. Karma generalizes the consequences, positive or negative, that ensue if an individual honors or fails to honor the requirements of Dharma. Buddhism, basing itself on Hindu metaphysics, takes Hinduism in the direction of religious psychiatry. First, it teaches that life is nothing but suffering, Buddhism's diagnosis of the human condition; and second, that the cause of suffering is desire. Hence, Buddhism prescribes the extinction of desire as a requisite therapy for the cessation of suffering. Since the ultimate goal for all sentient existence is the cessation of suffering, the realization of this Truth leading to self-managed psychotherapy through contemplation, and compassion for all sentient beings, are the contents of basic Buddhist norms.

Confucianism integrates a theory of a politically organized society with a theory of human nature in a comprehensive concept of Tao, the Way. Proper conduct of public and private affairs generalized through the Tao results in observance of custom, the practice of virtue and humanity. Islam bases its rules of conduct on the Holy Quran, the

Prophet's Sayings and Traditions, and the Consensus of the Moslem Community. Islamic norms come closer to real life than the norms of any other "advanced" society. They do not generalize what cannot be generalized.

Christianity strove for centuries to relate Roman and customary law to Christian moral teachings into what were supposed to be the coordinate realms of laws and morals. No other universal religion attempted to do that, and only Western civilization proposed distinguishing laws and morals. The idea that an action can be legally right but morally wrong, and vice versa, exists only in societies that were influenced by Christianity.

Patriarchal societies propound neither universal norms, nor proclaim universally valid values. Societies influenced by universal religions do. There is, however, no agreement between universal religions on which norms and values to accept and which to reject. All universal religions, with the possible exception of Islam, propound norms and profess beliefs in values that are not realizable at all, or only to some extent, or only by a meritorious few. None of them, including religions of patriarchal societies, propound norms and values that are compatible with the lifestyles and demands of contemporary societies. It seems that in the future, as is currently the case, no directing thrust, but at best a limited influence, will be exerted by religion in the business and governmental processes of contemporary societies.

A question of profound importance, though only speculative at present, is whether a universal, relevant, and influential religion or ideology can evolve to serve the needs of globally interconnected emerging societies and the individuals in them. Is it possible to determine universally acceptable norms and values of proscribed and prescribed behavior that are realistic, so that humans can live with and believe in them?

Most of an individual human life evolves through programmed biochemical processes. Only a small superstructure, the consciousness sphere, can function through self-determination. But even conscious decision making can hardly be said to be free of societal interactions. Yet at least some individual decision-making seems to be free. There is a difference of opinion as to which and to what extent certain decisions are freely made. Only some human behavior, to a greater or lesser extent, is controllable by conscious decision making. Therefore, the imputation of responsibility through conscious decision making falsifies facts in many cases, and is essentially a hit and miss game when figuring out reasons why somebody did or did not do something. It can be seen why this is so even in controlled criminal proceedings, where an effort is made to curb unbridled flights of fancy.

Thus, it can be taken for granted that the defendant, the lawyers, the jury members, and the judge are similar by virtue of being human beings. Hence, they should be able to judge each other's motives by analogical inference, a mode of reasoning used in law as well as in science. Let the defendant be charged with killing a man and let a proof for that be given to the jury. The issue to be decided at this point is whether the defendant killed with "malice aforethought." The prosecution claims that the defendant beat the victim with the intent to kill him and take his money. The defense claims that the defendant only intended to beat the hell out of the victim because the victim offended him. Who is the jury to believe? The motives claimed by the prosecution and the defense are mutually exclusive on psychological and logical grounds. This shows what a proof "beyond a reasonable doubt" is worth.

As one discerns prescribed and proscribed behavior in patriarchal societies, one can notice contours that emit signs of rationality and those that emit signs of irrationality. As one goes from the patriarchal societies to sophisticated ones, the contours of rationality are to be found mostly in governmental decision making processes. One would expect that the rational approach to managing relations between individuals and human collectives would tend to be on the increase. However, the opposite is often the case.

The conflicts between units within a government, social groups and countries generate a complexity that is destructive rather than enhancing to rational processes. The situation resembles in some ways turbulent and chaotic processes in nature rather than smooth operations of well-designed machines. Therefore, complete control of individual and collective behavior by means known at present is impossible. This means that neither force, nor punishment, nor persuasion, nor pressure, nor any combination of them is known to be effective in controlling the behavior of individuals and collectives by any reliable gauges. Social controls at present are punishment by physical means, shaming, disgracing and ostracizing; finally, the rewards through honors, economic benefits, within or without the social stratification confines. None of these have been shown to be always effective, or effective to some percentage point, or always ineffective. Some of them or a combination of them seem to work on some occasions but not on other occasions. This is just about all that can be inferred from past and present cases.

A graduated though definitive region of demarcation between industrialized and urbanized political setups and those that are not can be discerned by observing how one arrives at their policies. Those of individualized and urbanized states are arrived at through arguing and wrangling in politicized governmental bodies. Those of non-industrialized and non-urbanized states are arrived at through custom, tradition, and occasional political or military means. This means that the policies of industrialized and urbanized states are rationalized, though chaotic and sometimes confusing and inconsistent, due to political fighting and compromise. Those of non-industrialized and non-urbanized states tend to be less confusing and chaotic.

Every society prohibited killing and harming some humans while allowing killing and harming of others. Every society prohibited appropriating and damaging property of some while allowing taking and damaging properties of others. Thus, killing and harming members of one's own tribe was prohibited, while killing and harming members of alien tribes, especially in war, was not. The same goes for property. Attempts to control and restrict killing and injury, theft and robbery led to institutionalizing blood vengeance and war as a means of punishing violations of honor, body, and property.

The norms that evolved to protect life, honor, and property, and the institutions to pursue the violators, although not ubiquitous, are missing only in a few societies that we know existed on Earth. There is therefore no significant difference between "primitive" and "civilized" societies in this respect. A major difference arose with the introduction of explosives that depersonalized and denationalized blood vengeance and war. If quick evidence for this is needed, all one has to observe are the phenomena of artillery shelling, missile and aerial bombardment. There is literally nothing personal, tribal, or national in these. The victims of such can be saints or sinners, nationals of sundry nations, including those of the nation conducting such operations. Decisions to conduct artillery shelling, missile and aerial bombardment are made on an impersonal and denationalized basis. Executors of such decisions are essentially technician bureaucrats doing their job.

Some forms of sexual behavior were always taboo, and in general, human societies always strove to control sexual relations. There are wide differences between societies as to which sexual relations ought to be taboo and which not. In some contemporary societies, some minute differentiations as to which sexual relations to prohibit and which to allow beat even the medieval hairsplitting casuistry. In certain societies, incest, adultery, and fornication are taboo, but in others they are not. Homosexuality is in most societies taboo, though there are significant exceptions.

There is a ubiquitous feature that all but some contemporary societies exhibit: the overwhelming influence of religion to determine which forms of sexual behavior are to be taboo. In some contemporary societies, however, the role of the legislatures and the courts, and hence, of politics has become overwhelming. Indeed, lobbying to influence legislatures and courts in these matters shows how the sexual matters territory has been surrendered by religious claims, rather than affirmed. This means, of course, that sexual behavior is becoming more and more subject to the vicissitudes of governmental processes rather than some form of "natural law." The criminalization and decriminalization of sexual behavior, indeed of human behavior in general, and the absolutes of law and morals as they were known before the twentieth century, are eliminated either directly through legislation, or indirectly through judicial interpretation.

If one abandons the idea of an absolute source of the "moral imperative," of the Kantian or any other kind, it becomes difficult and perhaps impossible to determine the source of an individual's duty, obligation and responsibility. In particular cases, such as with contracts, the sources of obligation are the agreements of the contracting parties to honor their commitments, if they are given willingly and knowingly. It has been argued by the Social Contract theorists that the citizens of a state are likewise bound by a contract, the social contract. But whatever the ideological value of the doctrine may be, it is false to fact. To start, it is false to assume that people subjugated by a foreign power, or even by a junta of their own, ever agreed to a social contract authorizing their subjugation. Indeed, no social contract establishing a society ever existed, nor could ever exist such that those not yet born could be said to have been a party to the contract.

If one takes the empirical route and takes it as true to fact in the case of private law that the source of duty, obligation and responsibility is the contract, one is still left with the question as to what is the source in public law. That the state has to secure essential services to its citizens is obvious, as is obvious that it cannot render these without the reciprocal obligations of the citizens to the state. The Natural Law Doctrine, an absolutist doctrine, is compatible with this fact and can offer a plausible explanation as to why the state should be considered as a source of duty, obligation, and responsibility, with or without the Supreme Being claimed as the Ultimate Source of all moral and legal duties. However, neither the Natural Law Doctrine nor the presumption that the state has to secure essential services to its citizens are adequate to settle the limits to which the state can do this nor the extent of the duties the state can demand of its citizens and non-citizens. For example, on what grounds may the state prohibit, as it has almost always done, homosexuality and oblige its citizens and non-citizens to comply with the prohibition? Strangely enough, it is not the tribal state that claimed the right to criminalize and decriminalize behavior ad lib, but the modern "rule of law" state.

The sanctions for a delict on a contract in private law, and even in criminal law in patriarchal societies, have been compensatory and punitive damages. The sanctions for criminal behavior have been blood vengeance in some societies, opprobrium in all societies, and oftentimes death in many societies. The sanctions for violating treaties have sometimes been war, sometimes the cessation in interstate intercourse, and sometimes nothing. Regardless of what have been intended consequences of sanctions, which have been ostensibly to secure compliance, the results have been relatively effective in private law, ineffective in criminal law, and sometimes effective but sometimes ineffective in constitutional and administrative law.

Right and wrong, justice, equality, and fairness have been hypostatized in civilizations energized by Greco-Roman philosophers, Christianity, and Islam from Socrates to John Rawls. Battleships could be sunk with the weight of writings that purport to define them. None of these definitions provide a firmer ground for understanding and interpreting these concepts than the ordinary dictionary definitions. The emotive power loaded into them can hardly be exaggerated, although the operational use in interpersonal and societal relations is none. To convince oneself that this is indeed so, all one has to do is try to organize air traffic using them, and see what happens. In short, whenever technology and special training enter the scene, any attempts to operate through reliance on justice, fairness, equality and right and wrong will end in chaos and confusion, or worse. The rules capable of regulating and controlling sophisticated technological setups cannot be based on concepts alien and irrelevant to the setups. Thus, the competence that is essential for personnel in such setups necessitates grading and rating, in direct conflict with the requirements of equality and even justice and fairness. One could, of course, talk of justice and fairness in grading and rating, but then justice and fairness would amount to no more than rhetorical pleonasms purporting to cover ground already covered by competence in grading and rating. What this all means is that there are no sophisticated religions and ideologies loaded with the emotive power of justice, fairness, equality, right and wrong that are relevant and compatible with contemporary science, technology, and urban life.

There are several terms whose formulation goes back to ancient Greeks and whose analogs, though not equivalents, can be found in other civilizations as well. They are "honesty," "truth," and "character." All of them have been hypostatized in Western civilizations to mere symbols rather than working conceptions. First, the term "character" is a leftover of the pre-empirical psychology with at most a limited use in ordinary circumstances and no prospect of a definition capable of a positive or negative verification. Second, the term "honesty" denotes a definiendum that, when associated with a particular human, means that he or she never told a lie or cheated in the past, is not doing so at present, and will not do so in the future. Since there is no way of telling what somebody will tell or do in the future, even if through some miraculous means one could know everything that somebody said or did in the past, the claim that somebody is honest cannot be verified even in principle. It is rather certain on empirical grounds that everybody occasionally lied, or at least uttered a white lie. Lastly, an attempt to relativize the term to denote somebody who has usually told the truth and not cheated in the past makes the term "honesty" unusable in any serious enterprise.

Alfred Tarski raised the definition of truth to the level of mathematical rigor in contrast to the dictionary definitions and philosophical or theological disquisitions. Unfortunately, Tarski's definition of truth is unusable outside of logic, mathematics, and science. Nonetheless, the term "truth" has uses, though limited, in ordinary life; in the same way as its antonyms "untruth," "falsehood," and "lie" can have to register somebody's verbal behavior on some occasions. Exceptions to this are libel and slander to defame somebody's "character." In such cases, "lie" and its synonyms become reified to the weapons of hate, condemnation, and extralegal punishment.

Loyalty to the king and lese majesty, together with the de jure social stratification, were lifted ideologically by the American and French revolutions. However, the de facto social stratification through financial, educational, cultural, and societal positioning has increased in contemporary societies. This fact will make it ever more difficult for a majority of contemporary voters to cast a responsible vote in view of the fact that they do not know or understand the issues, nor are able to assess the qualifications of the candidates. The authority of the state, personified by the king, or depersonalized, as initially in the Roman state and most contemporary states, has always been presumed. But the Roman state and the United States of America stand out as the staunchest upholders of the principle. In the United States, this is so to the extent that not even the President is immune against the assertion of the Authority of the state against him or her. No citizen or combinations of citizens, no foreign state, implicitly or explicitly, are allowed to challenge the Authority of the United States without running the peril of being subdued and, if necessary, destroyed. Obvious though this fact is, not even Reinhold Niebuhr noticed it. The restraints to the exercise of the Authority of the United States against her residents are set by the Constitution. Those are, in particular, the right to peaceful protest against the government policies and enforcement actions, as well as the right to due process under the law.

There are no restraints against the policies of the United States Government involving foreign citizens and foreign countries unless such policies run counter to international law, as laid down in the treaties ratified by the United States. In general, contemporary states recognize no higher authority to which the authority of the state is subject. Therefore, the laws of the state that constitute the policy of the state are subject to no moral or legal review by another authority. Thus, the authority of the state is conceived as absolute by contemporary states. The question that now arises is how the policy of the state can be carried out. In domestic matters, the usual means is ultimately brute physical force, which may or may not be restrained by prudence. In foreign affairs, the usual means is also ultimately brute physical force, war, where prudence also may or may not be a restraint. Thus, in both domestic and foreign matters, the only restraints are the lack of power and the prudence of the ruling body in enforcing the policy. This means, again, the competence of the governing body.

If one subtracts ideologically motivated rhetoric of justice, equality, rights, and humanity, one finds that in any societal arrangement a cluster of interests determines the policy of the state and the manner of its execution. This cluster of interests falls into several categories: individual, group, national, and global. Politically organized societies have developed a variety of institutions to satisfy those interests. If one sets aside the Communist system that introduced radical institutional changes to fit its ideological needs, it can be said that institutions develop through the gradual growth within a given socio-economic setup to satisfy the needs of those they are expected to serve. Thus institutional changes occur in order to continue meeting particular needs, unless the institutions are unable to evolve through changes, in which case they become ossified and eventually die.

The fundamental need of every human is securing livelihood. It translates today into an economic interest. Today, as in the past, an individual's economic interest is pursued by herself or himself, aided or hampered by the family and other individuals. The role of the family and other individuals, however, is fading away in contemporary societies. So is the role of religious institutions that had provided for many individuals whose needs exceeded their and their families' resources. Today the state has taken over caring for human economic interests to a large extent. The basic means by which the state is doing this is by fostering or hampering economic activities through distribution and redistribution of income; the vehicles for this being taxation, entitlement payments, and welfare.

A major state interest has always been to protect and facilitate economic activity within its borders since a state's existence ultimately depends on its economic resources. However, religious and ideological interests have also been paramount and oftentimes took precedence over the economic interests.

How well or how poorly any of these interests faired depended on institutions developed by politically organized societies. There are, however, radical directional changes in all the societies on Earth due to the influence of the American Democracy and its variant of the Capitalist system. The tendencies are toward developing ponderous machineries for the decision making processes that are geared toward minimizing mistakes, which are endemic in areas where no effective decision making mechanisms can exist. The complicated mechanisms for checks and balances, rigorous supervision and control of personnel, decentralization when advantageous, and centralization when necessary are the hallmarks not only of the new methods of government, but also of the new ways of operating within the Capitalist system.

The American political system evolved originally with the design of a government that was to have separate and equal legislative, administrative and judicial branches. The underlying machinery for all the branches, however, began to evolve into more and more impersonal bureaucracies, with the requisite paperwork raised to ultimate levels of complexity and sophistication through the mechanization and computerization of operations. Strangely enough, business went in the same direction. Both prongs of American governmental and business design have kinks, but both work. In view of the demise of the monarchies and the unworkability of the Communist system, no other empirical alternative but the American model is left.

As one observes contemporary governmental and corporate institutions, it becomes noticeable that the essential attributes common to all of them are the efficacy in processing the input, the efficiency of the machinery, and the competence of the personnel running the machinery. Thus, the inputs in healthcare facilities are humans, beginning with prenatal care, through birth and lifelong care, and ending with death and burial, when the funeral industry takes over the processing. As one studies wheat or meat industries, similar types of processing are noticeable. Indeed, as one observes government or financial institutions, similar processing is evident, even though the inputs processed are human cases.

Since the institutional machineries set the norms of conduct and the manner of their enforcement, those being processed as well as those processing them are associated with the attributes of efficiency, efficacy, and competence. Since the systems themselves are to a decisive degree, if not completely, depersonalized, it can be seen that there is little room left for traditional values of the West like justice, prudence, fortitude, temperance, faith, hope, and charity, or their equivalents in non-Western societies. Since psychology also has no room for them they can be relegated to the religious systems to which they belong. The all-important question now is: are thy necessary at all, or can they be dispensed with? It is hard to believe that they can be dispensed with in view that they are indispensable in folk psychology wherever folk psychology is applied: in law and law practice, in medicine, in everyday life, even in psychiatry. The inescapable conclusion is that they are needed but not necessary in that there are alternative ways to accommodate the need.

The globalization of terrestrial societies is diluting and crisscrossing national interests to an unprecedented degree. Its influence in creating new norms and letting the old ones sink into desuetude can hardly be minimal. There is no religion or ideology in sight that has the vitality and appeal that could acquire a universal sway, rather than be just a regional phenomenon. This is a fact of contemporary life. Since the welfare state in one shape or another is also a fact of contemporary life, high levels of taxation to fund the welfare state, and other needs, is also a fact of life in the contemporary world. The use of war on a large scale à la Carl von Clausewitz as an instrument of foreign policy has receded, and could conceivably be abandoned altogether in view of the uncontrollability of outcome and concurrent ramifications. But, which way the war may go, the "heroic" virtues of "courage," "honor," etc. are not likely to play a significant role among the well-paid, well-trained, and professionalized military.

A general conclusion can now be reached: the norms of conduct, for individuals and collectives, are likely to be developed by legislatures, courts, administrative practice, and custom under the decisive influence of the need for efficiency, efficacy, and competence. The new norms of conduct are likely to develop incrementally as exigencies arise and as inherited norms fail to serve their intended purposes. The past belonged to kings and sometimes to charismatic leaders. The future belongs to technocrats and technicians, be they presidents, chief operating officers, administrators, production personnel, or clerks.

Modern democracies arose with the rise of industrialization and commercialization of Western societies and in opposition to the institutional setup of feudal societies. Strangely enough, their basic ideological tenets, an assortment of human rights, are derivatives from the feudal society. Even stranger is the fact that through the United Nations Charter and other transmitters, the idea of human rights, oftentimes in conflict with inherited tenets of non-Western societies, has been accepted throughout the world.

Initially, the medieval Church fought for and won the acceptance of a rather modest right: that a Christian slave had the right not to be sold to a non-Christian master. Then the feudal lords fought for and won an assortment of rights for themselves against the encroachment of absolutistically-minded kings. The right to a fair trial and due process was also hatched in Canon Law and judicial practice, as was the idea of the right of life. Therefore, the main thrust of modern democracies was directed toward securing equal treatment under the law, which implies a repudiation of de jure social stratification, but leaves intact de facto social stratification. The de jure equality, rather than the securing of rights in general, had a direct bearing on the rise of the middle classes and eventually the capitalist system, in that it removed the legal and political barriers for non-aristocrats.

The rise of the welfare state toward the end of the nineteenth and twentieth centuries became a necessity in view of the increasingly greater inability of individuals to secure means of livelihood on their own. The rise of the welfare state necessitated the need to fund its programs and secure funding through taxation. Taxation in turn caused the toning down of the de facto social stratification to the point where it is not an important factor in the management of human affairs.

The right to life generates a number of derivative rights. These can include the right to one's body and personality, the right to healthcare, and arguably the right to have children. If one accepts the right to one's body and personality, the issues of sexual conduct reduce to the non-infringement of the rights of others while exercising one's sexual freedom. This then includes the right to homosexuality. Indeed, this particular right does not burden the state with funding it, as the right to have children, for example, often does. The right to life, including the right o one's body and personality, militates against the claims of the state to restrict the right to life by the imposition of capital punishment and the restraints on the commission of suicide. These issues have not been addressed so far in a manner that promises meaningful decisional options.

The right to property is universally recognized as part and parcel of human existence. But there also always has been an interference with the right to property by the government through taxation, confiscation, or the inability of the state to prevent the taking of property through theft, robbery and fraud. These facts imply that the right to property reduces in practical operational terms to determining to what extent, and in which manner, an individual shall be allowed by the state to dispose freely of his or her property. The rights to receive healthcare and to procreate children are not controversial by themselves. They become difficult and perhaps impossible to deal with when the question of realistic limits and the related question of funding are raised. The burdens that the satisfaction of these rights will impose will become eventually unbearable not only in view of the tax burden but also in view of the exhaustion of the mineral, water, air, and space resources. Would the rationing of procreation and healthcare be the road to take, as some societies are already doing?

Equality of humans in biological and mental prowess is unachievable, and perhaps undesirable. Hence, equality of property is impossible. Equality in social status is not a realistic goal. A relatively equal treatment by governmental agencies is not only achievable, realistic, and desirable, but with a properly run bureaucratic machine inexpensive and simple.

Strangely enough, the right to free speech is more important to the society than to the individual, for it is a vital supervisory sensor against malfunctions of government and social machinery. It is, however, also open to wide-ranging uses for propaganda, blacklisting, defamation, and assorted falsifications. If one includes in media history oral and written literature, it can be stated without exaggeration that no society, past or present, has found a way to channel the media to their proper uses to the exclusion of the improper ones. The right to the freedom of religion in view that religious expression belongs to the broad concept of the media, has, just as other media, its proper and its improper uses. It also has its positive and negative sides, none of them, so far as is known, controllable in its uses.

An expensive right, propounded by many, is the right to education. Self-funded education is expensive to the individuals themselves, but funded education is expensive to others. The question now is: should the others be forced to pay for somebody else's education? Operation of contemporary machinery, technological and social, is impossible without a certain minimum education level for the whole population, higher levels for segments of the population, and ultimate levels for a small number of the population. The only question that remains then is the accessibility of higher education to be paid for by others. Implicit also at the educational level is grading. But grading is pyramidal, and always generates an elite. The majority of the population is therefore forced to pay for the creation of an educational elite. Relevant issues, therefore, are the choice of education subjects slated for public support, and the fairness of the grading methods. This means that the end product of properly administered education is the properly controlled creation of inequalities in areas selected for grading.

Prisons, until literally into the nineteenth and twentieth centuries, existed mainly for the purpose of torture and extracting ransom, without prison bureaucracies, and without the appurtenances of due process, human treatment of prisoners, rehabilitation, and the rest. Penal colonies resembling contemporary prison facilities were created in Tsarist Russia, the British Empire, and the United States. The inputs of contemporary prison facilities, a continuation of penal colonies, are the humans that had been processed as the output of the bureaucratic machinery, varying in levels of sophistication from country to country, or even within a given country. This output machinery creates a new segment in a population that is devoid de jure and de facto of many human rights for a term, or for life. There are a couple of issues that arise in connection with the punishment by incarceration and capital punishment as well. The first issue is the source of authority that allows a government to punish anybody for anything.

Attempts to justify punishment of human beings by human agents can be sorted into those that anchor the right to punish in a Supreme Being, as does the Natural Law doctrine; and those that anchor the right to punish in the "people," as do democracies and totalitarian government in a fashion. However, in view of inconsistencies in theory and in practice, either source of authority to punish is spurious. The second issue concerns proposals to look at punishment as a conduit for venting righteous indignation against offenders; proposals to look at punishment as a means of rehabilitating offenders; and finally, as a means of quarantining dangerous humans. None of these proposals, separately or jointly, can stand serious scrutiny.

The bureaucratic government machinery treats the offenders input as it processes them from arrest to indictment, trial, sentencing, and incarceration in a manner that is not much different from any processing machinery, such as in hospitals, slaughter houses, and even public schools. Thus, in more than one way the actual approach to punishment makes more sense than the so-called theories of punishment. In actual encounters, government faces assorted violators of the law and utilizes the bureaucratic machinery to dispose of them. It has the responsibility to dispose of law breakers in the same manner as the sanitation department has the responsibility to dispose of garbage, and it uses the means necessary to accomplish the disposition. The justification of any kind of punishment to accomplish the disposition, together with the means to be employed, remains an open question. Thus far, no credible answers have been given, nor are any in sight.

Rights always entail duties that are imposed on some, or all, to respect the claimant's rights which then in turn entail the duties to carry the financial and other burdens in order to honor the claimant's rights. The most burdensome of all rights is the right to wage war. The circumstances under which it is right to wage war is a moot question. Christian Church Fathers propounded that only a defensive war is justified. Contemporary international law takes the same position. Carl von Clausewitz propounded that war is nothing but a continuation of the state's policy by military means. There is no doubt but that contemporary sophisticated governments follow von Clausewitz' doctrine, while paying lip service to the Church Fathers and international law. But there is also no doubt but that the populations of both the victorious and the vanquished states have to suffer sacrifices of life and the destruction of property.

Reparations by the losers in the contemporary world are also illusory. They are as a rule economically weaker than the winners, unable to pay their own bills, and hence, in no position to pay reparations. The grotesqueness of the situation becomes crystal clear as one ponders the fact that the losers are always in the wrong if for no other reason than for waging a war they were bound to lose. The winners may be in the wrong if for no other reason than for winning a war that was not worth winning. The real victims are the populations of both the winning and the losing sides for the reason that they elected, were unable to remove, or simply suffered their governments. The question is thus wide open as to whether governments have the right to tax or otherwise burden the population for the purpose of waging war.

Where's the Future

The future of Earth depends on terrestrial and extraterrestrial factors. Extraterrestrial factors means a possible collision with another celestial body resulting in total or partial destruction of Earth and the quantum effects that would leave Earth uninhabitable, if not totally destroyed. If such events would occur, Earth and everything on it would not be missed in the universe. The destruction of Earth and its past and present inhabitants would not mean the Last Judgment in the universe, indeed not even in the solar system. Such terminology belongs to the language of religion, not the language of the universe. In case the Earth is not destroyed or devastated within a foreseeable future, one can glimpse opportunities for development in some directions and lack of such in others.

The usual approach to values like the beauty, the good, and the right is to launch disquisitions into what should be the case and ignore what at best could be the case. The unusual approach is undoubtedly a realistic one, the usual approach no more than good or bad fiction.

Cell manipulation and production of organisms are cases of biological programming. Its consequences are overcrowding of Earth and gradual depletion of soil and atmospheric resources. Biological, psychological, and sociological ramifications of such a situation are not hard to visualize. The upsides and the downsides of the cultural and environmental conditions of mankind are becoming further enhanced and further degraded. Impressions of what both would look like can then be left to individual imagination.

The control of Earth's overcrowding can only be effected by the control of population growth. The control over population growth can then be effected by genetic reprogramming or by infanticide, birth control, abortion, abstention, emigration from Earth, or a combination of such. A plausible guess is that most of them will be pursued. Since all of them are realistic, it makes sense to look into what is desirable, what is preferable, and what should not be the case.

Infanticide was practiced since ancient times to eliminate unwanted children. Abstention was also practiced to avoid unwanted pregnancies. Both, however, have only limited use as population growth controls. Birth control and abortion are presently in wide use, either foisted by the government upon unwilling populace, or freely used for reasons of personal convenience and family financial constraints. Birth control and abortion are effective when used; however, they had not been employed for any global population growth planning. Effective though birth control and abortion are, the question whether they should be used is wide open because of some undesirable consequences. Without changing the original bioprogram, it is clear that there are a host of ramifications in any attempt to restrain en masse human procreative proclivities. Thus, on this account, one should rule out birth control and abortion as desirable methods of controlling the population growth.

Ethically unacceptable ramifications of foisted birth control and abortion are linked with human rights, and hence cannot be given a diminished or preferred status vis-à-vis other human rights. So the right to have children is a basic human right and has to be recognized as such, provided that it is not accorded at the expense of other human rights.

The control of population growth by bioengineering appears to be a distinct possibility. However, its consequences, if it is ever used, are so far reaching that no sensible discussion on the basis of empirical evidence is possible. One thing, though, is certain: the original design of the Programmer would be drastically altered for better or worse by it. But, if one believes that the original Programmer's design is sacrosanct, then one would have to reject any tampering with it on religious and moral grounds.

Emigration from Earth into outer space seems to offer the most promising future for mankind, especially so since it requires no drastic departures from established traditions in local cultures and the emerging global civilization. Since emigration to outer space is in no conflict with the customary ways of managing human affairs, it is likely that it will combine with other ways of controlling population growth and perhaps soften up the harshness of at least forced population growth controls.

Traditional legal and moral values were always challenged in ethnic, cultural, and religious conflicts. However, until Friedrich Nietzsche and Karl Marx, none of the challenges ever came, within the same tradition, by self-reflection. Nietzsche pointed to the conflict between Christian values and what he believed should be the right values. So Nietzsche proposed the transvaluation of all values, meaning traditional Western values, in favor of the right values. Marx pointed to the contradictions within what he considered the Capitalist System, in essence the socio-economic system of the West. Distorted spin-offs of the Nietzschean and Marxian ideas are the demised National Socialism and Communism. Both are stark testimony to the power of ideological brews, cooked and served by political and philosophical amateurs. Indeed, any radical religious and ideological clamor for instant justice eventually leads to disastrous consequences if unchecked.

Traditional values of the West are challenged today from several sources. One challenge is due to the incompatibility of traditional Western values and contemporary science. This challenge creates a conflict within the civilization itself. Other challenges come from non-Western sources, in particular from violent convulsions in Islamic countries.

Since non-Western collectives are so far the receivers, it is fitting to target the analytic laser on the West as the dispenser of the cultural products to the world at large. To avoid getting caught in the web of unsolved mind-body problems when dealing with cultural products, it will suffice to use "mind" and "body" in the ordinary language usage. The medieval moral thought, evolving through enormous efforts of Western philosophers and theologians, reached its peak in the Thomistic synthesis. The system of St. Thomas Aquinas integrated the threads of political, philosophical, and theological thought into a relatively consistent set of ideas. This enabled the West, in spite of some internal contradictions, to form a relatively coherent outlook on the universe, the earth, and mankind.

Major breakthroughs in science and technology, beginning with the Renaissance and still going on full force, destroyed the Thomistic system's capability to confront successfully modernity. The Protestant efforts to vie with these problems, by Paul Tillich and Reinhold Niebuhr, must be deemed unsuccessful. The outcome of all this is that there are many conflicting voices in the West that are not what they claim to be. Some have called the situation bankruptcy. Actually, it is worse than moral bankruptcy, for it implies that a basic legal and moral orientation is impossible, that the very concepts of right and wrong are meaningless, and that the religious sources of Western law and moral teachings dried up.

As an illustration, one may ponder the profound problems associated with capital punishment and killing in war. The Church Fathers faced them head on. To justify capital punishment, the Church Fathers proposed that society has the right to self-defense, just as individuals do. But the claim to self-defense had to be genuine. Hence, an elaborate due process mechanism was developed in the bosom of the Church to ensure that only the guilty ones were punished by death. Likewise, Church Fathers thought that killing in a defensive war, and only a defensive war, is justifiable. Many did not accept the musings of the Church Fathers. At that time heretics, and later some Christian sects, believed that the biblical prohibition of killing was absolute, and hence no justification for killing of criminals or enemies in war was permissible. It is true that the Church's doctrine was never fully complied with either in peace or in war. All the same, however, the position of Western society was clear. Therefore, the essential ingredients for rationality and sanity were there.

It is true that today major and minor countries profess adherence to the Church's medieval and current doctrine. Thus modern states, at least the developed ones, make extraordinary and expensive efforts to ensure that only the guilty ones are sentenced to die. The same states, however, practice artillery shelling and aerial bombardment as a permissible means of implementing their policies. The victims of such policies are not even known ahead of time to the policies' makers, let alone proven guilty of even being the enemies. The ingredients for irrationality and insanity are all there, even if the policies are intended to combat "evil empires." The grotesqueness is enhanced if the high beam is trained on the policies of "rogue states."

It seems virtually certain that, short of cataclysmic events, the future of mankind will evolve in the same directions as at present with enhancement in some areas and diminution in others. The biological needs for food, sex, and shelter will certainly not be met for the majority of humans. This means that securing basic human rights and property can amount to no more than mere rhetoric for the vast majority of mankind, at least for the near future. The reason for this is simply that no effective methods of social control are known at present, nor are any in sight, that would make possible the implementation of any plan intended to secure these rights.

As of now, the only methods of social control in human collectives are conditioning, corporal punishment, and restricting access to opportunities. Corporal punishment that includes methods of inflicting pain and death is at best effective in very limited ways. Even ardent advocates of corporal punishment would not stake their life and property, because of a "credible threat" of corporal punishment to potential offenders. Social conditioning, with perhaps the restriction of access to opportunities, is more effective than other methods of social control but is by no means adequate by itself. Since there is no crime that is not associated with psychiatric disorders, psychotropic drugs, short of genetic reprogramming, show considerable promise for controlling individual behavior if used in conjunction with other forms of psychiatric treatment. Since there are no effective means of social control, there can be no effective ways of securing life and property at present or in the near future. Indeed, with the improvements in technological means for delivering services in contemporary societies, there are also improvements of capabilities to inflict loss and damage by mischief makers.

There is no historical evidence for the existence of regulated relations between human collectives by means of enforceable agreements, balance of power, or by the imposition of the superior power. So a semi-chaos and partially ordered relation disturbed by intermittent war and disorder are the historical record of the game. Thus, Thomas Hobbes' idea that "might is right" proves sometimes right and sometimes wrong. There is no known way to regulate relations between global collectives so that "might is right" is always, or at least usually, wrong. It is clear that there is something in the original programming of human beings that does not allow them to accept the idea that "might is right." Yet, there is no escaping the reality that might appears very often to be all that there is to the right.

The best contemporary educational programs have proved adequate for training technocrats who can function fairly competently as well-drilled players. None have proved themselves adequate to prepare leaders that can escalate their collectives to higher levels of relationships. A dimension of what has been known in every culture as spirituality and wisdom is missing in the contemporary world. But that is exactly what is needed in order to lead to higher levels of social existence. Contemporary education is impotent to create it; contemporary religions that once had it lost it. There is the need in the original program of humans for spirituality and wisdom that would energize contemporary men and women to higher levels of existence, with inward and outward growth and without the chartable limits. Short of it, the future world is "more of the same."

It will be instructive to construct a catalogue of future prospects on earth and beyond. To begin with, one has to take into account available land on earth: flora, fauna, the atmosphere, mineral resources, water, and finally the possibility of replenishing some of these from outer space. Then, one has to consider human capabilities to face challenges of nonhuman factors as well as those created by human activity.

The nonhuman factors can be quickly dealt with. Everything on Earth, including the Earth itself, is finite in size. Hence only a population of limited size can be housed, fed, and have access to water and air on Earth. Therefore, the only endeavor to prevent the catastrophe on Earth has to come from human actions and inactions.

As for human factors, the ones to contemplate are the Nation State, the United Nations, the capitalist economic system, religions, ideologies, extraterrestrial outer space, and their possible control mechanisms.

Preceding the American and French revolutions, neither representative nor political democracy existed. Parenthetically, representative and political democracies are not equivalent to the direct democracy of ancient Athens, nor do representative and political democracies entail economic and social equality.

The current constellation of human societies grouped in Nation States is arrayed in the United Nations, affiliated under the leadership of the United States of America as a successor to the defunct League of Nations. The Charter of the United Nations is its organic law. Since almost all states of the world are members of the United Nations, it is taken for granted that the tenets propounded by the Charter are universal. The United Nations has on paper what is needed for ordinary operations: the deliberative units, the court, and the administrative unit. It plays, however, a relatively minor role in the world due to its inability to deal effectively with the problems that brought it into existence.

The capitalist economic system was and is the most effective and efficient system for the production of goods and services. Not even Karl Marx and his followers challenged this fact. To wit: nobody challenged the production side of the system. The criticism of the system was and is focused on the distribution side of the system.

This system developed in Europe and North America with the expectation that natural resources and skilled labor would be available for use if needed capital is available. These were the real reasons for its success. Slogans like "growth," "creativity," "productivity," and "innovation" were meant for the hoi polloi. The exhaustion of natural resources, the degradation of the ecosystem and, in the words of Joseph Schumpeter, "creative destruction"—these were the real consequences of capitalism's unstoppable thrust.

"Creative destruction" created mechanized processes of production, thereby eliminating less productive enterprises and the need for unskilled labor. All of this created a permanent load on the economy and society in the form of human throwaways. The situation is poignantly evident in the political campaign slogan "I will fight for you." Fighting means war, for the throwaways' poor, a permanent war.

Primitive religions occur usually in institutionalized forms in patriarchal societies of tribal or extended family types of social organization. These societies are undergoing convulsive transformations on the road to Westernization and are not a factor, but a problem in the world.

Henotheism is a type of religion that professes a belief in the god of a particular, usually ethnic, group. Henotheistic types of historically known religions are not a factor in the contemporary world. Of particular interest in this context are the now defunct National Socialism and Fascism. Their leaders were accorded hero worship that resonated like the worship of a henotheistic god. Their demise, however, is an instructive example. In times of social stress, cries for accountability and justice zero in not only on the majority but on the ostracized minority.

The concept of universal religion applies to Hinduism, Buddhism, Christianity, and Islam. Buddhism is an atheistic religion, and communism is an atheistic ideology. An interesting question is why communism failed in the Soviet Union but not in China. There is an essential difference that is immediately noticeable between China and the Soviet Union. China and Japan have exhibited an ability to absorb and integrate cultural imports unmatched in history.

Thus, China imported Hinduism and Buddhism, Western communism, and capitalism; absorbed them and integrated them into its culture. Czarist Russia was not able to fully absorb either Western communism or Western capitalism. Hinduism and Christianity are in a state of anemia and barely able to perform vital societal roles they did in the past.

Rapid industrialization and urbanization have torn apart families and the local community, basic social units in the past. Some form of social reintegration through some form of religion or ideology will be needed to restore social and psychological health throughout the world.

Islamic countries are in a somewhat analogous position with Czarist Russia, which has been in convulsions since Peter the Great. Since Peter the Great, Russia has oscillated between efforts to westernize or to preserve its Byzantine heritage with its spiritual and cultural components. Islamic countries are in a state of confrontation with the American style of Western civilization that has influenced the entire world from head to toe: from software, computers, cars and movies to bikini swimsuits and thong underwear.

In the past, Islam energized Arab caliphates, the Ottoman and Mogul empires, and developed its own ways of social control under the Sharia and the governments' organic law of the Byzantine and other vintages. Hence, it is much easier for Islamic countries to accept Western modes of government than the Western "values" that are incompatible with Islamic ones.

The United Nations offers the best extant framework to work on the problems of preserving the ecosystem on Earth, secure peace in the world, and develop needed deep sea and outer space legal controls. Therefore, the question is whether the great powers that control the United Nations can work together to realize the principles spelled out in the United Nations Charter.

Glossary

Algorithm: Semblance of algorithmic reasoning is as old as natural language. It was already used by ancient Hindu, Greek and Arabic mathematicians.

The basic idea of what an algorithm is expected to deliver is illustrated in criminal trials. What has to be decided in criminal trials are questions of permissible evidence, the rules of law and procedure. These are inputs. The verdict of guilty or not guilty is the output. If the jury cannot reach the verdict, it is deemed hung.

The formal sciences (logic, mathematics, and the computation theory) use a mix of natural language and symbols to make rigorously sanitized formal language. Hence, the language used to express assumptions and the inferential machinery is free of vagueness and ambiguity. Thus, the assumptions processed by the inferential machinery are the input in formal sciences. The output are the deduced conclusions.

If the output expresses a true statement it is called *decidable*, or equivalently, *computable*; and if it expresses a false statement it too is *decidable*, or equivalently *computable*.

However, the formalization of logic, mathematics, and the theory of computation quickly led to Kurt Goedel's proof, and later Alan Turing's proof, that there are statements in formal sciences that are undecidable or uncomputable. This means that there are, for example, statements of arithmetic that cannot be proved true or false. This also means that the problems formulated by such statements are unsolvable.

The general conclusion is thus that no algorithm exists that can deliver the output "true" or "false" for every input in formal sciences.

Amnesia: Indicates by the dictionary definition greater or lesser loss of memory. However, the dictionary definition of amnesia is meaningless unless one knows what memory is. Furthermore, psychiatric diagnosis considers that dementia, aphasia, Alzheimer's, etc., also indicate the loss of memory. It so appears that memory is a grab-bag from which all of these ailments and defects issue forth.

The failure of psychiatric diagnoses to pin down these mentalistic concepts as well as many others is due to not paying attention to linguistic conundrums involved. There are words in natural languages that are used to name single entities and classes of entities. There are also words serving other purposes. For present purpose only the meaning of words are of interest.

First, naming words in formal sciences (logic, mathematics, and the theory of computation) are introduced by formal definitions. For example, "two," "zwei," "dva" name in the written form the same entity which is the number 2. There are in this case three different words that name the same abstract entity.

Second, let the words "sun," "sonne," "sunce" name in different languages the same concrete entity. Any attempt to identify this entity will have to be based on sensory perception with or without aid of instruments.

Third, let the words "consciousness," "bewusstsein," and "svest" name the same abstract entity. There is no known way of identifying this abstract entity. The only evidence that warrants the claim that people know what they are talking about when they use the word "consciousness" is the explanation proposed by the American linguist Noam Chomsky. Chomsky's explanation is simple and conclusive.

For example, "Peter had consciousness for breakfast" satisfies the phonetic rules of English language if a native speaker says it. It is also grammatically correct. But it is a semantic anomaly in any natural language.

However, "Peter lost consciousness due to the accident" satisfies the phonetic rules of the English language if uttered by a native speaker of English. It also satisfies the grammatical rules of English, and also semantic rules of any natural language. But it satisfies the phonetic and grammatical rules of English only.

A general conclusion can now be reached. Names, simple or compound, occurring in natural or formal languages, in spoken or written form, can name concrete or abstract entities.

Gottlob Frege introduced the term "sense" as an attribute of "name" and the term "reference" as an attribute of the entity named.

The term "meaning" can be introduced to denote the properly related complex of sense and reference. For example, if one understands the word "two" and can identify its reference, then one understands the meaning of the number 2.

These requirements are satisfied in formal sciences. But, it should be added that even in formal sciences not all definitions are conclusive to the ultimate degree of conclusiveness.

The sense of a name in natural sciences cannot be rigorously fixed but the reference being tangible increases the intelligibility of the meaning to a high degree. There are so far no conclusive ways in social sciences for determining the reference although the determination of sense is no insurmountable problem.

Amnesia, aphasia, Alzheimer's, and *dementia* are excellent illustrations for the problems of definition in social sciences.

> **Amnesia**: Indicates greater or lesser loss of memory by the dictionary definition. However, the dictionary definition of amnesia is meaningless unless one knows what memory is.
>
> **Aphasia**: The dictionary defines it as the language disturbance that impairs the ability to speak, write or comprehend the meaning of spoken or written words. Does that mean that such an individual has the memory of words that is not using it? It is impossible at present to find out if she or he cannot write or speak.
>
> **Dementia**: The dictionary definition of dementia is that this disorder of the mind impairs perception and memory. That an individual has a perception of an object can be determined by questioning. But that such an individual has an impairment of memory or indeed has a mind cannot be inferred by the fact that such an individual perceived an object.

Aquinas, Saint Thomas (1225-1274): Italian Catholic priest and important medieval philosopher. St. Thomas employed Aristotle's philosophical writings to construct a systematic exposition of his philosophy. His primary works include *Summa Theologica* and *Summa Contra Gentiles*. In these works, St. Thomas provided a systematic exposition of the Christian doctrines and a trenchant defense against their critics. St. Thomas' system of philosophy was proclaimed by Pope Leo XIII in 1879 the official philosophy of the Catholic Church.

Aristotle (384-322 B.C.): Greek philosopher. Aristotle's literary works were not preserved. What were preserved were the notes of his lectures compiled by his students. By these notes Aristotle became the teacher of the West to the present day and for centuries, of the Islamic world as well.

Aśoka (304-232 B.C.): Also known as Aśoka the Great. Emperor of India from 269 to 232 BC. Aśoka embraced Buddhism and was instrumental in the spread of Buddhism throughout Asia.

Brahman-Atman: Also known as Brahman or Brahma. In the Vedanta school of Indian philosophy, the ultimate reality or *atman* (self) from which the universe emanates.

Cardinal number: Indicates the number of members in a set of numbers. If a set has no member, its cardinal number is 0; if it has one member, its cardinal number is 1; if it has as many members of the set of natural numbers (0,1,2,3 . . .), its cardinal number is *ALEPH NULL*; if it has as many members as the set of real numbers, its cardinal number is *ALEPH ONE*; and so on.

Charge (electron): An electron has a negative charge. Its antiparticle is the positron.

Chomsky, Noam (1928–): American linguist who revolutionized grammatical research and created what is called "generative grammar." Chomsky is also active writing on political subjects that are not within his expertise.

Church, Alonzo (1903–1995): American mathematical logician who made important contributions to mathematical logic and the theory of computation.

Clausewitz, Karl von (1780–1831): Prussian General and writer on military strategy. His book *On War* had a major impact on military strategy and foreign policy thinking.

Conservation Laws: Consist of the following: Conservation of angular momentum; conservation of charge; conservation of energy; conservation of linear momentum; conservation of potential energy; conservation of strangeness (a property of some elementary particles that decay slower in the process of energy release).

Dementia: Vide supra at *Amnesia*.

Deterministic Law: Usually refers to a law of Newtonian Physics. The Newtonian law implies that there are no exceptions to the description of events it governs.

Einstein, Albert (1879–1955): American theoretical physicist. He was a major contributor to the development of quantum mechanics, one of the branches of contemporary physics. Einstein was the sole originator of relativity theory, the other branch of contemporary physics.

Electro-magnetism: The relation between electric and magnetic phenomena is the subject of electro-magnetism.

Entropy: The second law of thermodynamics, the field of natural sciences dealing with the conversion of energy to heat and heat to energy. The name was introduced by Rudolf Clausius (1822–1888). It covers huge territory. As a factor in changes in the atmosphere, it shows the directions of the heat flow. It is also a factor in changes of biological processes.

Euclidean geometry: Composed by Euclid (300 B.C.) with the title *Elements*. It became the textbook of geometry to the present time.

The *Elements* was presented in the form of definitions and postulates to serve for the deductions of theorems. The *Elements* start with the definitions and postulates. There are five postulates, the fifth of which is the crucial one. It says: given a straight line l and a point A that is not on l implies that one can draw only one straight line through A parallel to l. Mathematicians for centuries tried to infer the parallel (fifth) postulate from the other four unsuccessfully.

During the 19th century successful challenges to Euclid's fifth postulate were made my Nikolai Lobachevsky, Janos Bolyai and Bernhard Riemann that led to the development of non-Euclidean geometries.

The importance of non-Euclidean geometries became manifest with the rise of Relativity Theory, whose geometry is the geometry of outer space. The important point is that the Euclid's Fifth Postulate was changed only at a crucial point: the manifold on which the angles of a triangle are measured.

The angles of a triangle sum to 180 degrees on a plane Euclidean surface. In the hyperbolic geometry (Bolyai and Lobachevsky) the angles of a triangle sum to less than 180 degrees. And in elliptical geometry (Riemann), on the surface of a sphere the angles of a triangle sum to more than 180 degrees.

Extrapolation: Used in science as well as in everyday life without cognizance of the basis on which it rests.

Let S be an arbitrary set of entities, finite or infinite. Suppose that S is a finite set. Then, it can be closed or open. If S is closed, then, in principle every member of it can be checked to determine what kind of attributes it has. In such a case no extrapolation is needed since inferences about the members of the set will be based on the concrete observation.

If the set is open the inspection of all of its members may not be possible. Therefore, the observed attributes on a subset of the set may be used to extrapolate about the attributes of the unobserved members of the whole set (past, present, and future). This sort of inspection and extrapolation is employed in experimental sciences. The insistence that the experiment be repeatable to reconfirm the previous results is in effect recognition that this kind of extrapolating can never be conclusive.

Let the set S be infinite and open or closed. Then the extrapolating can be conclusive depending on the type of members of the set. For example: suppose that one wants to find out whether the whole infinite set of even natural numbers is divisible by 2 without remainders. One can start by checking this subset of the members of the set of even natural numbers: 2/2, 4/2, 6/2, 8/2, 10/2, 12/2, 14/2, etc. The conclusion by extrapolation is that the whole set of even natural numbers is divisible by 2.

Suppose now that one wants to inspect a finite and open set of known earthquakes to find out how often they occur, where they occur and the time of occurrence. Once the inspection is complete it will become obvious that no extrapolation will deliver the information as to when and where the next earthquake will occur. Extrapolation therefore cannot be used in this case.

Goedel, Kurt: (1906–1928): Austrian-American mathematician. Major achievements of Goedel were the proof of the predicate logic completeness; the proof that the formal system of Principia Mathematica (Bertrand Russell's and Alfred North Whitehead's work) and related systems are incomplete; and finally, that the continuum hypothesis and the axiom of choice cannot be disproved.

Grammatical function words: Used to join components of words (conjunctions, prepositions, interjections).

Gravitational Constant: Occurs in the statement of Newton's Third Law. It is represented by the symbol g and its value at sea level is 32 feet/second.

Hegel, Georg W.F. (1770–1831): German philosopher influential in the rise of political ideologies of the nineteenth and twentieth centuries. The story is that the left wing of Hegel's philosophy was led by Karl Marx, who originated the Communist ideology, and the right wing was represented by the National Socialist ideology. The climax of the conflict between them was the battle of Stalingrad, resulting in the defeat of the Hegelian Right by the Hegelian Left.

Henotheism: World religions are classified into polytheistic, henotheistic, and monotheistic. The belief in many gods and goddesses is the identifying characteristic of polytheism. Belief in one god, usually of an ethnic group, without denying the existence of other groups' gods, is the identifying characteristic of henotheism.

Hobbes, Thomas (1588–1679): English philosopher. His main work, *Leviathan*, is a classic of political and legal philosophy.

Jakobson, Roman (1896–1982): Russian-American member of the Prague school of linguistics known for his pioneering works on aphasia.

Kant, Immanuel (1724–1804): German philosopher, a major figure in the modern Western philosophy known as Transcendental Idealism. According to Kant, there are two classes of entities: the class of phenomena and the class of noumena.

The class of phenomena is accessible to sensory experience. This is the class of concrete entities that are referred to by their names.

The class of noumena transcends sensory experience. This is the class of abstract entities; the reference of abstract entities, being accessible only to "pure reason." They are space, time, justice, beauty, etc.

Magnetic constant: The permeability of free space defined as the ratio of magnetic flux density and external field strength.

Maxwell's equations: James Clerk Maxwell (1831–1879) formulated these equations to unify the fields of electricity and magnetism and, in effect, made them the foundation of the field of electromagnetism.

Metalanguage: Suppose one wants to learn German using English. In such a case German is the "object language" and English is the "metalanguage." Or, suppose one wants to learn what is meant by the Newtonian formula "F=mxa." In this case the object language is "F=mxa" and the English explanation that this formula means, "force is the product of mass and acceleration," is the metalanguage.

Metaphysical constant: U is introduced to name the ultimate source of all existence and thus to provide a point at which cosmological speculations, supported by incomplete empirical evidence, and even the intuitions of the mystics can converge.

There are at present a couple of competing cosmological theories, the prevalent being the so-called "standard model of the universe" and the competing being the "endless universe." There are then the intuitions of the Hindu sages who called the ultimate source of all existence Brahman-Atman, and the mystics of the universal religions (Hinduism, Christianity, Islam) that called it the "unknown." The "unknown," they claimed, is unnamable and indefinable, accessible only through religious experience.

Minkowsky, Hermann (1864–1909): Russian-German mathematician who formulated the concept of the space-time to create the geometry for the Special Relativity.

Marx, Karl (1818–1883): German economist and philosopher. His main works, *The Capital* and *The Communist Manifesto*, contain the tenets for both the socialist and communist ideologies.

Model: Purports to name classes of concrete and abstract objects (of physics, economics, everyday life and so on). The references associated with the sense of the name "model" are determinable only by linguistic context or sensory experience; except within the field of mathematics. In other words, for non-mathematical objects only dictionary definitions are possible.

An example of mathematical modeling is the analytic geometry. It was René Descartes (1596–1650) who showed how to represent geometrical entities arithmetically and conversely. A simple example of how this is done in two-dimensional space is this: A geometrical point is defined in Euclidean geometry as an entity that has no magnitude. Its arithmetical representation is (X,Y), that is, the ordered pair in the language of arithmetic. But the ordered pair (X,Y) also represents a geometrical point (X,Y) of the Cartesian coordinate system.

Niebuhr, Reinhold (1892–1971): Prominent American theologian and political thinker.

Natural law doctrine: Originated in ancient Greece and was propounded in the Roman Empire by the Stoic philosophers. It influenced the development of jurisprudence in the Roman Empire, the Middle Ages and the twentieth century.

The fundamental tenet of the natural law doctrine is the belief that there are absolute normative principles that take precedence over other legal and moral principles. Furthermore, the Natural Law Doctrine holds that human beings by their very constitution know what they are. In other words claiming that one does not know what is right and what is wrong is no defense.

Newtonian mechanics: Also known as Classical Mechanics, still the macrophysics in the terrestrial realm by virtue of being validated through the Special Relativity for speeds less than the speed of light.

Nietzsche, Friedrich Wilhelm (1844–1900): German poet and philosopher.

Object language: Vide supra at metalanguage.

Perelman, Chaïm (1912–1984): Polish-Belgian philosopher of jurisprudence and a writer on rhetoric.

Ontology: A field of philosophy whose study is the nature and purpose of existence.

Ordinal number: A member of an ordered set of entities. Ordinal numbers in natural languages are (in English) *first, second, third, fourth*, etc.

Planck constant: Expresses the relation between quantum energy and frequency.

Plato (427–347): Ancient Greek philosopher who communicated philosophy using imaginary dialogues between his teacher, Socrates, and interlocutors. Plato created the dialogue form in literature and he was the grand master in communicating philosophy, which in Ancient Greece comprised science, mathematics and liberal arts.

The core of Platonic doctrine is visualized ideas that are representations of classes of concrete and abstract entities. The relationship between the visualized ideas and associated entities is termed "participation of particulars in universals."

Contemporary philosophy of mathematics and the foundations of mathematics are almost all Platonic. It is obvious why this is so if one realizes that the visualized ideas are abstract mathematical entities.

Popper, Karl (1902–1994): Austrian-British philosopher who distinguished himself more as a philosopher of science than a political philosopher.

Positron: The antiparticle of the electron.

Private law: The lore of law is sorted into private and public law. Public law consists of constitutional, criminal, and other laws governing the relations between individuals and the state. Private law regulates relations between individuals, including companies and corporations.

Probabilistic laws: Non-deterministic laws formulated by means of the probability calculus.

Prosodic features: Gesticular features of natural language.

Public law: Vide Supra at *Private law*.

Quantum mechanics: Major divisions of modern physics are macrophysics, which consists of variants of the relativity theory, and microphysics, which consists of quantum mechanics and thermodynamics.

Quine, Willard Van Orman (1908–2000): American philosopher who successfully applied mathematical logic for research and analysis of logical and semantic characteristics of natural language.

Ranke, Leopold von (1795–1886): German historian who set the standards for writing objective history based on evidence, free of emotive language and propaganda.

Rawls, John (1921–2000): American philosopher known for his work *A Theory of Justice*, an apologia of liberal democracy. Rawls was a major influence in legal, moral and political philosophy.

Recursive (or inductive) definitions: Computability or recursive theory was formulated by Kurt Goedel and Stephen Cole Kleene using recursive definitions of functions. It was then proved that these functions are equivalent to computable functions.

For example, the recursive or inductive definition of the set of natural numbers by means of the function "successor of" is:

1. *0* is a natural number (the initial condition);
2. For every *n*, if *n* is a natural number, then the successor of n is a natural number (the inductive condition);
3. Nothing else is a natural number (the closure condition).

Relativity theory: Formulated by Albert Einstein purportedly by replacing the conceptions of absolute space and time of Newtonian physics with relativistic conceptions. It is unnecessary to explain what Newton took to be space and time. It will suffice to quote a sentence from a recent book to see what the situation is in contemporary physics:

> "The first thing that can be said is we do not honestly know the true nature of space and time." [*On Space and Time*, ed. Shahn Majid, Cambridge University Press, 2008, p. XI]

What Einstein and Hermann Minkowski actually did was to provide the mathematical framework that combined the names "space" and "time" into the compound name "space-time" and introduced the metric tensor (the basic concept of tensor calculus) to handle distances in the many dimensional spaces. Tensor calculus and Riemannian geometry supplied the relativity theory with the mathematics needed to handle effectively local and outer space distances.

Riemannian geometry: Vide supra at *Euclidean geometry*.

Rousseau, Jean-Jacques (1712–1778): Swiss philosopher who exerted enormous influence on political and educational thinkers. Rousseau's chief works are *The Social Contract, or Principles of Political Right*, and *Emile: or, On Education*.

Saussure, Ferdinand de (1857–1913): Swiss linguist. Saussure, Roman Jakobson and Noam Chomsky are the leading figures of twentieth-century linguistics.

Schumpeter, Joseph (1883–1950): Austrian-American economist who, in his work *Capitalism, Socialism and Democracy*, elaborated the term "creative destruction" introduced by German economist Werner Sombart.

Spengler, Oswald (1880–1936): German historian and philosopher of history. Spengler claimed that the course of life valid for biological organisms is also valid for civilizations. They too start with the birth, go through the maturity, and then die. These views were presented by Spengler in his work *Decline of the West*. In this work he also claims that Western civilization is in the stage of irreversible decline.

Tarski, Alfred (1901–1983): Polish-American logician and mathematician best known for his work on model theory and for introducing the terms "object language" and "metalanguage."

Thermodynamics: The field of physics and chemistry whose subjects of study are heat and energy.

Tillich, Paul Johannes (1886–1965): German-American theologian and philosopher. In his *Systematic Theology,* Tillich tried to accomplish for twentieth-century Protestantism what St. Thomas did for medieval Catholicism.

Toynbee, Arnold Joseph (1889–1975): British historian. In his work *A Study of History,* Toynbee countered attempts to find irreversible patterns in history. Toynbee thought that the course of history is determined by challenge and the response to challenge. In other words, he did not accept the idea of inevitability in history.

Withering of state: The idea of some Communist ideologies that once an egalitarian and proletarian society is established there will not be any need for government to maintain law and order; hence, the state will wither.